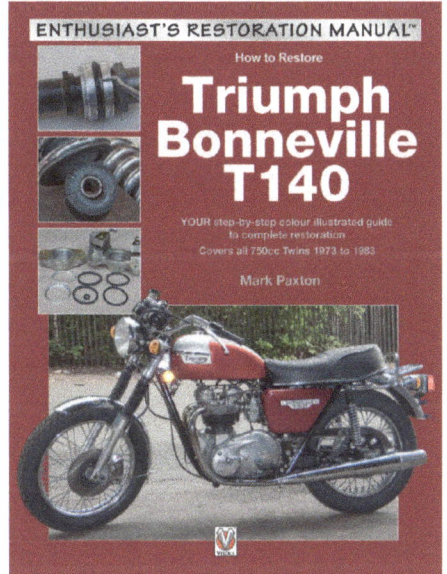

Other great books from Veloce –

1½-litre GP Racing 1961-1965 (Whitelock)
AC Two-litre Saloons & Buckland Sportscars (Archibald)
Alfa Romeo 155/156/147 Competition Touring Cars (Collins)
Alfa Romeo Giulia Coupé GT & GTA (Tipler)
Alfa Romeo Montreal – The dream car that came true (Taylor)
Alfa Romeo Montreal – The Essential Companion (Classic Reprint of 500 copies) (Taylor)
Alfa Tipo 33 (McDonough & Collins)
Alpine & Renault – The Development of the Revolutionary Turbo F1 Car 1968 to 1979 (Smith)
Alpine & Renault – The Sports Prototypes 1963 to 1969 (Smith)
Alpine & Renault – The Sports Prototypes 1973 to 1978 (Smith)
An Austin Anthology (Stringer)
An Austin Anthology II (Stringer)
An English Car Designer Abroad (Birtwhistle)
An Incredible Journey (Falls & Reisch)
Anatomy of the Classic Mini (Huthert & Ely)
Anatomy of the Works Minis (Moylan)
Armstrong-Siddeley (Smith)
Art Deco and British Car Design (Down)
Autodrome (Collins & Ireland)
Automotive A-Z, Lane's Dictionary of Automotive Terms (Lane)
Automotive Mascots (Kay & Springate)
Bahamas Speed Weeks, The (O'Neil)
Bentley Continental, Corniche and Azure (Bennett)
Bentley MkVI, Rolls-Royce Silver Wraith, Dawn & Cloud/ Bentley R & S-Series (Nutland)
Bluebird CN7 (Stevens)
BMC, The Cars of (Robson)
BMC Competitions Department Secrets (Turner, Chambers & Browning)
BMW 5-Series (Cranswick)
BMW Z-Cars (Taylor)
BMW Classic 5 Series 1972 to 2003 (Cranswick)
BMW M3 & M4, The complete history of these ultimate driving machines (Robson)
British at Indianapolis, The (Wagstaff)
British Cars, The Complete Catalogue of, 1895-1975 (Culshaw & Horrobin)
BRM – A Mechanic's Tale (Salmon)
BRM V16 (Ludvigsen)
Bugatti – The 8-cylinder Touring Cars 1920-34 (Price & Arbey)
Bugatti Type 40 (Price)
Bugatti 46/50 Updated Edition (Price & Arbey)
Bugatti T44 & T49 (Price & Arbey)
Bugatti 57 2nd Edition (Price)
Bugatti Type 57 Grand Prix – A Celebration (Tomlinson)
Camaro 1967-81, Cranswick on (Cranswick)
Caravan or Motorhome Habitation Check, Do Your Own (Shephard)
Caravan, Improve & Modify Your (Porter)
Caravans, The Illustrated History 1919-1959 (Jenkinson)
Caravans, The Illustrated History From 1960 (Jenkinson)
Carrera Panamericana, La (Tipler)
Car-tastrophes – 80 automotive atrocities from the past 20 years (Honest John, Fowler)
Chevrolet Corvette (Starkey)
Chrysler 300 – America's Most Powerful Car 2nd Edition (Ackerson)
Chrysler PT Cruiser (Ackerson)
Citroën DS (Bobbitt)
Classic British Car Electrical Systems (Astley)
Classic Engines, Modern Fuel: The Problems, the Solutions (Ireland)
Cobra – The Real Thing! (Legate)
Cobra, The last Shelby – My times with Carroll Shelby (Theodore)
Competition Car Aerodynamics 3rd Edition (McBeath)
Competition Car Composites A Practical Handbook (Revised 2nd Edition) (McBeath)
Cool Recipes & Camping Hacks for VW Campers (Richards)
Concept Cars, How to illustrate and design – New 2nd Edition (Dewey)
Cortina – Ford's Bestseller (Robson)
Cosworth – The Search for Power (6th edition) (Robson)
Coventry Climax Racing Engines (Hammill)
Daily Mirror 1970 World Cup Rally 40, The (Robson)
Daimler SP250 New Edition (Long)
Datsun Fairlady Roadster to 280ZX – The Z-Car Story (Long)
Dino – The V6 Ferrari (Long)
Dodge Challenger & Plymouth Barracuda (Grist)
Dodge Charger – Enduring Thunder (Ackerson)
Dodge Dynamite! (Grist)
Dodge Viper (Zatz)
Dorset from the Sea – The Jurassic Coast from Lyme Regis to Old Harry Rocks photographed from its best viewpoint (also Souvenir Edition) (Belasco)
Draw & Paint Cars – How to (Gardiner)
Drive on the Wild Side, A – 20 Extreme Driving Adventures From Around the World (Weaver)
Driven – An Elegy to Cars, Roads & Motorsport (Aston)
Dune Buggy, Building A – The Essential Manual (Shakespeare)
Dune Buggy Files (Hale)
Dune Buggy Handbook (Hale)
East German Motor Vehicles in Pictures (Suhr/Weinreich)
Electric Cars – The expert Q&A guide (Henshaw)
Essential Guide to Driving in Europe, The (Parish)
Fast Ladies – Female Racing Drivers 1888 to 1970 (Bouzanquet)
Fate of the Sleeping Beauties, The (op de Weegh/Hottendorff/op de Weegh)
Ferrari 288 GTO, The Book of the (Sackey)
Ferrari 333 SP (O'Neil)
Fiat & Abarth 124 Spider & Coupé (Tipler)
Fiat & Abarth 500 & 600 – 2nd Edition (Bobbitt)
Fiat in Motorsport (Bagnall)
Fiats, Great Small (Ward)
Ford Cleveland 335-Series V8 engine 1970 to 1982 – The Essential Source Book (Hammill)
Ford F100/F150 Pick-up 1948-1996 (Ackerson)
Ford F150 Pick-up 1997-2005 (Ackerson)
Ford Focus WRC (Robson)
Ford GT – Then, and Now (Streather)
Ford GT40 (Legate)
Ford GT40 Anthology – A unique compilation of stories about these most iconic cars (Allen & Endeacott)
Ford Maverick and Mercury Comet 1970-77, Cranswick on (Cranswick)
Ford Midsize Muscle – Fairlane, Torino & Ranchero (Cranswick)
Ford Model Y (Roberts)
Ford Mustang II & Pinto 1970 to 80 (Cranswick)
Ford Small Block V8 Racing Engines 1962-1970 – The Essential Source Book (Hammill)
Ford Thunderbird From 1954, The Book of the (Long)
Ford versus Ferrari – The battle for supremacy at Le Mans 1966 (Starkey)
Formula 1 - The Knowledge 2nd Edition (Hayhoe)
Formula 1 All The Races - The First 1000 (Smith)
Formula One – The Real Score? (Harvey)
Formula 5000 Motor Racing, Back then ... and back now (Lawson)
Forza Minardi! (Vigar)
France: the essential guide for car enthusiasts – 200 things for the car enthusiast to see and do (Parish)
The Good, the Mad and the Ugly ... not to mention Jeremy Clarkson (Dron)
Grand Prix Ferrari – The Years of Enzo Ferrari's Power, 1948-1980 (Pritchard)
Grand Prix Ford – DFV-powered Formula 1 Cars (Robson)
Great British Rally, The (Robson)
GT – The World's Best GT Cars 1953-73 (Dawson)
Hillclimbing & Sprinting – The Essential Manual (Short & Wilkinson)
Honda NSX (Long)
Honda S2000, The Book of The (Long)
Immortal Austin Seven (Morgan)
Inside the Rolls-Royce & Bentley Styling Department – 1971 to 2001 (Hull)
Intermeccanica – The Story of the Prancing Bull (McCredie & Reisner)
Jaguar - All the Cars (4th Edition) (Thorley)
Jaguar from the shop floor (Martin)
Jaguar E-type Factory and Private Competition Cars (Griffiths)
Jaguar, The Rise of (Price)
Jaguar XJ 220 – The Inside Story (Moreton)
Jaguar XJ-S, The Book of the (Long)
Jeep CJ (Ackerson)
Jeep Wrangler (Ackerson)
The Jowett Jupiter – The car that leaped to fame (Nankivell)
Karmann-Ghia Coupé & Convertible (Bobbitt)
Kris Meeke – Intercontinental Rally Challenge Champion (McBride)
KTM X-Bow (Pathmanathan)
Lamborghini Miura Bible, The (Sackey)
Lamborghini Murciélago, The book of the (Pathmanathan)
Lamborghini Urraco, The Book of the (Landsem)
Lancia 037 (Collins)
Lancia Delta HF Integrale (Blaettel & Wagner)
Lancia Delta Integrale (Collins)
Land Rover Design - 70 years of success (Hull)
Land Rover Emergency Vehicles (Taylor)
Land Rover Series III Reborn (Porter)
Land Rover, The Half-ton Military (Cook)
Land Rovers in British Military Service – coil sprung models 1970 to 2007 (Taylor)
Lea-Francis Story, The (Price)
Le Mans Panoramic (Ireland)
Lexus Story, The (Long)
Little book of microcars, the (Quellin)
Little book of smart, the – New Edition (Jackson)
Lola – The Illustrated History (1957-1977) (Starkey)
Lola – All the Sports Racing & Single-seater Racing Cars 1978-1997 (Starkey)
Lola T70 – The Racing History & Individual Chassis Record – 4th Edition (Starkey)
Lotus 18 Colin Chapman's U-turn (Whitelock)
Lotus 49 (Oliver)
Lotus Elan and +2 Source Book (Vale)
Lotus Elite – Colin Chapman's first GT Car (Vale)
Lotus Evora – Speed and Style (Tippler)
Making a Morgan (Hensing)
Marketingmobiles, The Wonderful Wacky World of (Hale)
Maserati 250F In Focus (Pritchard)
Mazda MX-5/Miata 1.6 Enthusiast's Workshop Manual (Grainger & Shoemark)
Mazda MX-5/Miata 1.8 Enthusiast's Workshop Manual (Grainger & Shoemark)
Mazda MX-5 Miata, The book of the – The 'Mk1' NA-series 1988 to 1997 (Long)
Mazda MX-5 Miata, The book of the – The 'Mk2' NB-series 1997 to 2004 (Long)
Mazda MX-5 Miata Roadster (Long)
Mazda Rotary-engined Cars (Cranswick)
Maximum Mini (Booij)
Meet the English (Bowie)
Mercedes-Benz SL – R230 series 2001 to 2011 (Long)
Mercedes-Benz SL – W113-series 1963-1971 (Long)
Mercedes-Benz SL & SLC – 107-series 1971-1989 (Long)
Mercedes-Benz SLK – R170 series 1996-2004 (Long)
Mercedes-Benz SLK – R171 series 2004-2011 (Long)
Mercedes-Benz W123-series – All models 1976 to 1986 (Long)
Mercedes-Benz W124 series – 1984-1997 (Long)
Mercedes G-Wagen (Long)
MG, Made in Abingdon (Frampton)
MGA (Price Williams)
MGB & MGB GT– Expert Guide (Auto-doc Series) (Williams)
MGB Electrical Systems Updated & Revised Edition (Astley)
MGB – The Illustrated History, Updated Fourth Edition (Wood & Burrell)
The MGC GTS Lightweights (Morys)
Micro Caravans (Jenkinson)
Micro Trucks (Mort)
Microcars at Large! (Quellin)
Mini Cooper – The Real Thing! (Tipler)
Mini Minor to Asia Minor (West)
Mitsubishi Lancer Evo, The Road Car & WRC Story (Long)
Montlhéry, The Story of the Paris Autodrome (Boddy)
MOPAR Muscle – Barracuda, Dart & Valiant 1960-1980 (Cranswick)
Morgan Maverick (Lawrence)
Morgan 3 Wheeler – back to the future!, The (Dron)
Morris Minor, 70 Years on the Road (Newell)
Motor Movies – The Posters! (Veysey)
Motor Racing – Reflections of a Lost Era (Carter)
Motor Racing – The Pursuit of Victory 1930-1962 (Carter)
Motor Racing – The Pursuit of Victory 1963-1972 (Wyatt/ Sears)
Motor Racing Heroes – The Stories of 100 Greats (Newman)
Motorhomes, The Illustrated History (Jenkinson)
Motorsport In colour, 1950s (Wainwright)
N.A.R.T. – A concise history of the North American Racing Team 1957 to 1983 (O'Neil)
Nissan 300ZX & 350Z – The Z-Car Story (Long)
Nissan GT-R Supercar: Born to race (Gorodji)
Nissan – The GTP & Group C Racecars 1984-1993 (Starkey)
Northeast American Sports Car Races 1950-1959 (O'Neil)
Nothing Runs – Misadventures in the Classic, Collectable & Exotic Car Biz (Slutsky)
Patina Volkswagen, How to Build a (Walker)
Patina Volkswagens (Walker)
Pass the Theory and Practical Driving Tests (Gibson & Hoole)
Pontiac Firebird – New 3rd Edition (Cranswick)
Porsche, Cranswick on (Cranswick)
Porsche 356 (2nd Edition) (Long)
Porsche 356, The Ultimate Book of the (Long)
Porsche 908 (Födisch, Neßhöver, Roßbach, Schwarz & Roßbach)
Porsche 911 Carrera – The Last of the Evolution (Corlett)
Porsche 911R, RS & RSR, 4th Edition (Starkey)
Porsche 911 SC, Clusker
Porsche 911, The Book of the (Long)
Porsche 911 – The Definitive History 1963-1971 (Long)
Porsche 911 – The Definitive History 1971-1977 (Long)
Porsche 911 – The Definitive History 1977-1987 (Long)
Porsche 911 – The Definitive History 1987-1997 (Long)
Porsche 911 – The Definitive History 1997-2004 (Long)
Porsche 911 – The Definitive History 2004-2012 (Long)
Porsche 911, The Ultimate Book of the Air-cooled (Long)
Porsche – The Racing 914s (Smith)
Porsche 911SC 'Super Carrera' – The Essential Companion (Streather)
Porsche 914 & 914-6: The Definitive History of the Road & Competition Cars (Long)
Porsche 924 (Long)
The Porsche 924 Carreras – evolution to excellence (Smith)
Porsche 928 (Long)
Porsche 930 to 935: The Turbo Porsches (Starkey)
Porsche 944 (Long)
Porsche 964, 993 & 996 Data Plate Code Breaker (Streather)
Porsche 993 'King Of Porsche' – The Essential Companion (Streather)
Porsche 996 'Supreme Porsche' – The Essential Companion (Streather)
Porsche 997 2004-2012 'Porsche Excellence' – The Essential Companion (Streather)
Porsche Boxster – The 986 series 1996-2004 (Long)
Porsche Boxster & Cayman – The 987 series (2004-2013) (Long)
Porsche Racing Cars – 1953 to 1975 (Long)
Porsche Racing Cars – 1976 to 2005 (Long)
Porsche - Silver Steeds (Smith)
Porsche – The Rally Story (Meredith)
Porsche: Three Generations of Genius (Meredith)
Powered by Porsche (Smith)
Preston Tucker & Others (Linde)
RAC Rally Action! (Gardiner)
Racing Camaros (Holmes)
Racing Colours – Motor Racing Compositions 1908-2009 (Newman)
Racing Mustangs – An International Photographic History 1964-1986 (Holmes)
Rallye Sport Fords: The Inside Story (Moreton)
Renewable Energy Home Handbook, The (Porter)
Roads with a View – England's greatest views and how to find them by road (Corfield)
Rolls-Royce Silver Shadow/Bentley T Series Corniche & Camargue – Revised & Enlarged Edition (Bobbitt)
Rolls-Royce Silver Spirit, Silver Spur & Bentley Mulsanne 2nd Edition (Bobbitt)
Rover P4 (Bobbitt)
Runways & Racers (O'Neil)
Russian Motor Vehicles – Soviet Limousines 1930-2003 (Kelly)
Russian Motor Vehicles – The Czarist Period 1784 to 1917 (Kelly)
RX-7 – Mazda's Rotary Engine Sportscar (Updated & Revised New Edition) (Long)
Sauber-Mercedes – The Group C Racecars 1985-1991 (Starkey)
Schlumpf – The intrigue behind the most beautiful car collection in the world (Op de Weegh & Op de Weegh)
Singer Story: Cars, Commercial Vehicles, Bicycles & Motorcycle (Atkinson)
Sleeping Beauties USA – abandoned classic cars & trucks (Marek)
SM – Citroën's Maserati-engined Supercar (Long & Claverol)
So, You want to be a Racing Driver? (Fahy)
Speedway – Auto racing's ghost tracks (Collins & Ireland)
Sprite Caravans, The Story of (Jenkinson)
Standard Motor Company, The Book of the (Robson)
Steve Hole's Kit Car Cornucopia – Cars, Companies, Stories, Facts & Figures: the UK's kit car scene since 1949 (Hole)
Subaru Impreza: The Road Car And WRC Story (Long)
Supercar, How to Build your own (Thompson)
Tales from the Toolbox (Oliver)
Tatra – The Legacy of Hans Ledwinka, Updated & Enlarged Collector's Edition of 1500 copies (Margolius & Henry)
Taxi! The Story of the 'London' Taxicab (Bobbitt)
This Day in Automotive History (Corey)
To Boldly Go – twenty six vehicle designs that dared to be different (Hull)
Toleman Story, The (Hilton)
Toyota Celica & Supra, The Book of Toyota's Sports Coupés (Long)
Toyota MR2 Coupés & Spyders (Long)
Triumph & Standard Cars 1945 to 1984 (Warrington)
Triumph Cars – The Complete Story (new 3rd edition) (Robson)
Triumph TR6 (Kimberley)
Two Summers – The Mercedes-Benz W196R Racing Car (Ackerson)
TWR Story, The – Group A (Hughes & Scott)
TWR's Le Mans Winning Jaguars (Starkey)
Unraced (Collins)
Volkswagen Bus Book, The (Bobbitt)
Volkswagen Bus or Van to Camper, How to Convert (Porter)
Volkswagen Type 3, The book of the – Concept, Design, International Production Models & Development (Glen)
Volkswagen Type 4, 411 and 412 (Cranswick)
Volkswagens of the World (Glen)
VW Beetle Cabriolet – The full story of the convertible Beetle (Bobbitt)
VW Beetle – The Car of the 20th Century (Copping)
VW Bus – 40 Years of Splitties, Bays & Wedges (Copping)
VW Bus Book, The (Bobbitt)
VW Golf: Five Generations of Fun (Copping & Cservenka)
VW – The Air-cooled Era (Copping)
VW T5 Camper Conversion Manual (Porter)
VW Campers (Copping)
Volvo Estate, The (Hollebone)
You & Your Jaguar XK/XKR – Buying, Enjoying, Maintaining, Modifying – New Edition (Thorley)
Which Oil? – Choosing the right oils & greases for your antique, vintage, veteran, classic or collector car (Michell)
Works MGs, The (Allison & Browning)
Works Minis, The Last (Purves & Brenchley)
Works Rally Mechanic (Moylan)

www.veloce.co.uk

First published in November 2017, reprinted February 2022 by Veloce Publishing Limited, Veloce House, Parkway Farm Business Park, Middle Farm Way, Poundbury, Dorchester DT1 3AR, England. Fax 01305 250479 / e-mail info@veloce.co.uk / web www.veloce.co.uk or www.velocebooks.com.
ISBN: 978-1-787111-49-3 UPC: 6-36847-01149-9
© 2017 & 2022 Mark Paxton and Veloce Publishing. All rights reserved. With the exception of quoting brief passages for the purpose of review, no part of this publication may be recorded, reproduced or transmitted by any means, including photocopying, without the written permission of Veloce Publishing Ltd. Throughout this book logos, model names and designations, etc, have been used for the purposes of identification, illustration and decoration. Such names are the property of the trademark holder as this is not an official publication. Readers with ideas for automotive books, or books on other transport or related hobby subjects, are invited to write to the editorial director of Veloce Publishing at the above address.
British Library Cataloguing in Publication Data – A catalogue record for this book is available from the British Library. Typesetting, design and page make-up all by Veloce Publishing Ltd on Apple Mac. Printed and bound by CPI Group (UK) Ltd, Croydon, CR0 4YY.

ENTHUSIAST'S RESTORATION MANUAL™

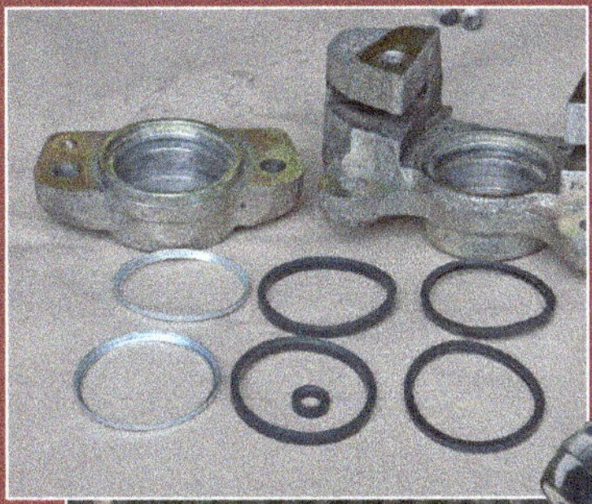

How to Restore

Triumph Bonneville T140

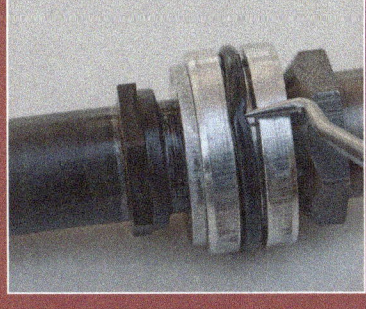

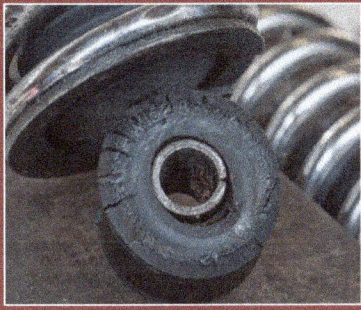

YOUR step-by-step colour illustrated guide to complete restoration

Covers all 750cc Twins 1973 to 1983

Mark Paxton

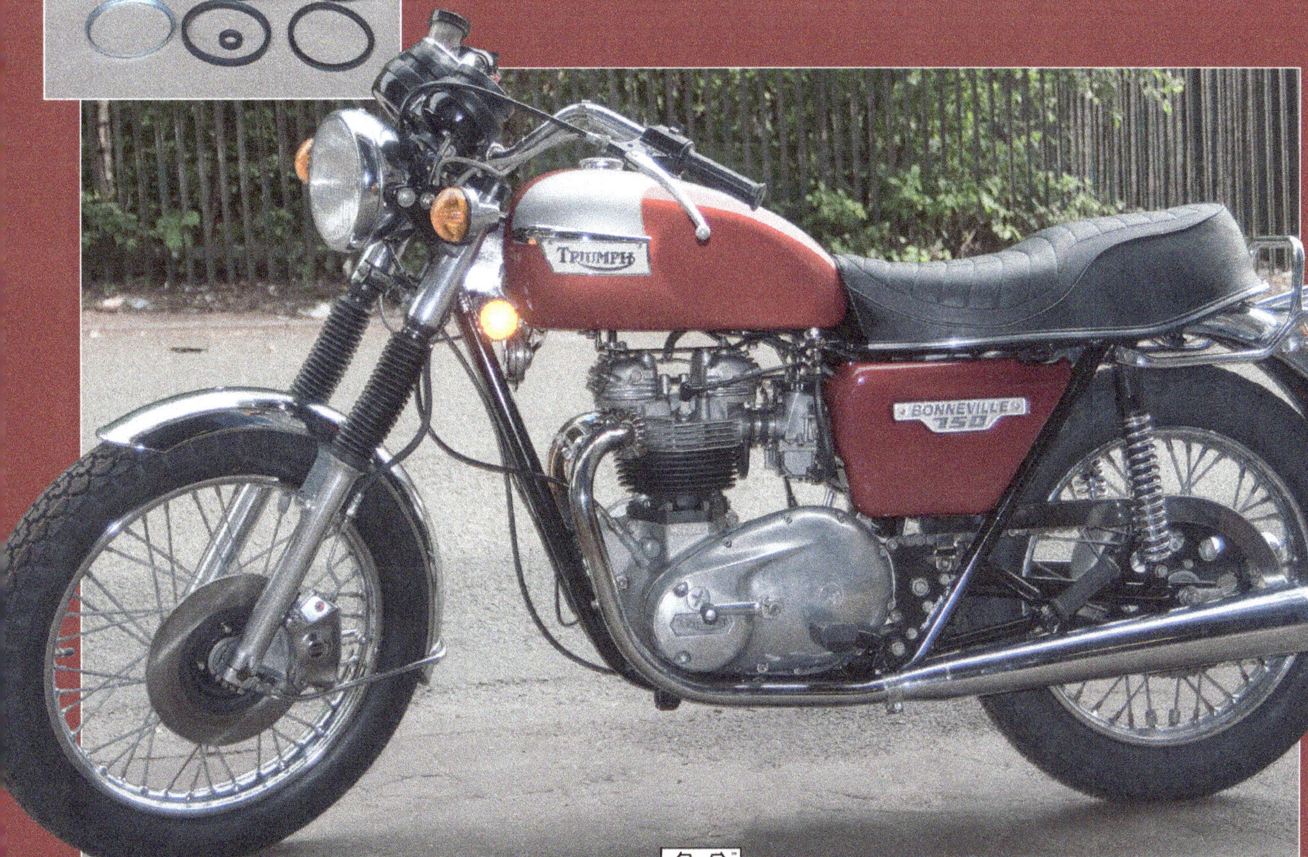

VELOCE

Contents

Introduction **6**
Why a Triumph T140? 6
Using this book 7
Join the Club 8
Project management 8
Parts 9
Hand tools 9
Big tools 10
Measuring tools 10
Workshop safety 10
Disclaimer 11
Thanks 11

Chapter 1 The plan **12**
Initial inspection 12
Run it or strip it? 12
General issues 13
Strip sequence 14

Chapter 2 The frame **16**
Preparation 16
Cracks and damage 18
Powder coating 18
Painting 18
Swinging arm 20
Oil filter 21
Centre and side stands 21
Battery box and coil mounting
   plate 22
Handlebars 22
Steering head bearings 23
Steering head lock 27

Chapter 3 Forks and shocks **28**
Forks 28
   Removal 28
   Stripping and rebuilding 29
   Refitting 30
Rear suspension units 34
Standard Girling 34
   Gas Girling 34
   Marzocchi Strada 34
   Marzocchi Strada AG 35
   Paioli 35
   Shock rebuild 35

Chapter 4 Brakes **38**
Master cylinders 38
Calipers 43
Discs 48
Brake pads 48
Rear drums 48
Brake hoses and pipes 49
Brake light switches 49
Brake fluid 50
Brake bleeding 50

Chapter 5 Engine and gearbox ... **51**
Top end strip 51
Top end inspection/overhaul 55
Threads and studs 55
Rocker boxes 56
Cylinder head 58
Pistons and rings 63
Barrel 64

Engine removal 67
Primary side 68
Timing side 74
Gearbox strip 75
   Inner cover 76
   Outer cover 76
Bottom end strip 83
Tachometer drive 83
   Oil pressure unit 83
   Splitting the cases 83
Crankshaft 85
Camshafts 85
Gearbox high gear 87
Case cleaning 87
Reassembly 87
   Bottom end rebuild 87
   Gearbox 90
   Primary side 92
   Timing side 93
   Barrels and pistons 94
   Cylinder head 94

Chapter 6 Fuel and exhaust **96**
Fuel tank 96
Fuel taps 97
Tank sealant 97
Carburettors 99
   Amal Concentric Mk1 99
   Amal Concentric Mk2 103
   Bing 109
Carb cleaning 109
   Manual 109

CONTENTS

Ultrasonic	109
Carb finishing	109
Fuel and additives	109
Exhausts	109

Chapter 7 Wheels 111
Wheel rebuilding 111
Mag wheels 111
Tyres 112
Wheel bearings 112
Speedo drive 113

Chapter 8 Paint 114
Decision time 114
Paint choice 114
What you need 115
Paint removal 115
Etch 115
Filler 116
Workshop preparation 117
Masking up 117
Primer 118
Prime again 118
Base coats 118

Two-tone 118
Candy 120
Lacquer 120
Finishing touches 121
Painting fibreglass 122

Chapter 9 Trim and brightwork 124
Transfers/decals 124
Seat replacement/recovering 125
Instruments 129
Chrome parts 130
   Mudguards 131
   Headlamp/fork shrouds 131
   Grab rail/rack 132
General plating 132
Polishing 133
Rubber parts 134
Badges 135
Rear light/reflectors/lenses 135

Chapter 10 Electrics 136
Loom 136
Wiring repairs 139

Alternator and voltage control ... 140
   Zener diode 140
   Rectifier 140
Ignition systems 141
   Electronic ignition 141
   Points and condenser 142
   Plugs and leads 143
   Coils 143
Battery 144
Lights and indicators 144
Horn 145
Switches 146
Electric start 147

Chapter 11 Putting it back together 148
Problems 151
Electrics 151
Carbs 151
Clutch 151
Oil 152
Was it worth it? 153

Index 160

Introduction

"It's a ride that every motorcyclist should enjoy at least once."
Cycle Magazine USA 1978

"The bike that created its own legend."
Triumph Factory Advertising

WHY A TRIUMPH T140?

Forty years is a considerable period of time in anyone's motoring career and, having passed that milestone, I realised that despite having owned, ridden, worked on, and restored dozens of bikes in that period, my experience with British iron was very sketchy indeed. A brief flirtation with a Tiger Cub and a C15 were pretty much it as far as ownership went; and I really didn't imagine that they represented the highlight of our great engineering traditions. So how would I address this yawning chasm in my motorcycling experience? Opinion

0.1 Admittedly, there isn't a T140 in the picture, but the sentiment is as strong today as when NVT placed this advert in American magazines, and reflected my excitement at getting my hands on an iconic British bike.

0.2 This 1976 advert managed to dilute some of the legendary kudos, with the addition of the Easy Rider model at the bottom, which was sadly just a collection of rebadged Bianchi moped bits.

INTRODUCTION

0.3 The brochure picture of how my project would have looked originally. The fork gaiters were not ultimately fitted to US market bikes, although the Canadians apparently had them.

0.4 The UK version had a larger tank and lower bars, to suit our riding style. They resulted in the (in)famous 'sack of potatoes' riding position, regularly criticised by road testers of the day.

0.5 The single-carb Tiger was easier to keep in tune, and better on fuel – a sensible choice today; unfortunately I couldn't find one.

seemed to point me squarely in one direction: a Bonneville (or even better a Tiger) 750 seemed to be regarded as the best introduction to this new world. Projects were still out there, although prices were rising rapidly, so I would have to get a move on. Spares were freely available, and the bike was universally regarded as a bona fide classic.

The result was the purchase of an American import – or should I say repatriation: a 1979 T140E, chosen, as it seemed to represent the sort of bike an eager-but-inexperienced Brit bike enthusiast like myself would be drawn to for restoration. It was in one piece, pretty much original, but clearly it had not run for several years. The realities of the restoration, good and bad, are laid out in the pages that follow.

USING THIS BOOK

This volume makes no pretence at being a comprehensive workshop guide for every variant of Triumph 750 made. Instead, it seeks to cover common restoration tasks in some detail, and should be used in close conjunction with the Factory Workshop Manual. A copy of the Parts Book for your particular model is also essential. These volumes are freely available online, or in paper form from your chosen parts suppliers.

The contents are a guide to renovation, restoration and repair; if you intend a complete rebuild to the original specifications, then you will also have to invest in factory brochures, and copies of magazine road tests, to ensure that everything is period correct. Even then, Meriden was well known for bolting bits onto bikes as they came to hand, especially during the sit-in, and the long slow decline at the end, so factory mix-and-matching is far from uncommon. I have also tried to include information on the most common options, even when they were not directly covered during my rebuild.

All the work in the book has been done by me, as there really is no substitute for personal experience to get a feel for the effort, skill or nerve-jangling frustration involved in the restoration process. The

HOW TO RESTORE Triumph Bonneville T140

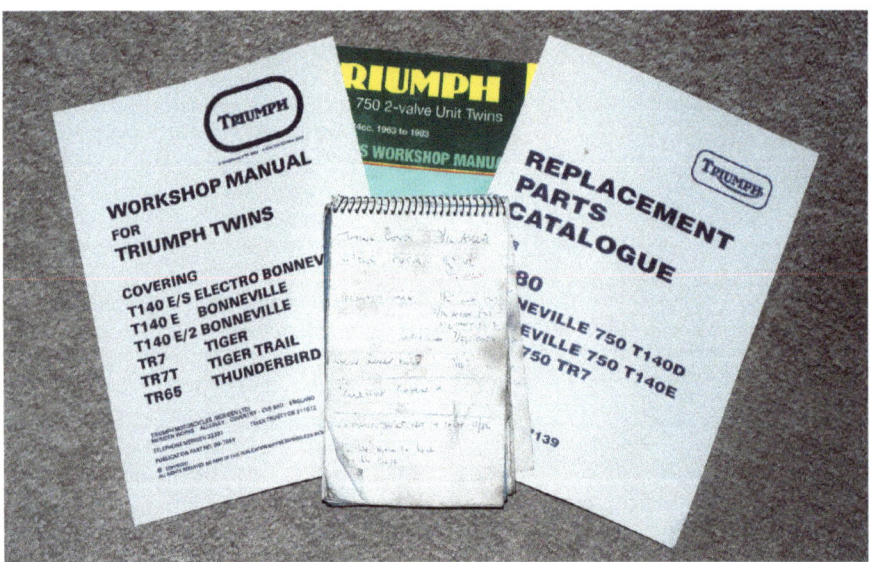

0.6 The Factory Workshop Manual and Parts Book are essential. An aftermarket manual is also a good idea. A notebook and pen would help keep track of all the bits that didn't come up to scratch.

0.7 If you are going for a factory-fresh restoration, then you will have to collect as much period material as you can find, although even that was not always entirely accurate.

techniques outlined may differ from those advocated by some, or even the factory itself, at times, but they all definitely work. Opinions are also expressed, which are the result of my personal experience; others may disagree with them – just choose the solution that best matches your needs, preferences, and pocket. It is written by an enthusiast, hopefully for other enthusiasts, using the sort of tools and under the conditions that a home restorer would be familiar with.

It is probably best to read through the book completely to get a feel for the whole endeavour, before diving into your particular restoration. On occasions, the picture sequence may show parts, fitted or missing, which may or may not have been discussed up to that point (probably simply due to a personal whim during the work), so it is best to stick with the sequence laid out in the text, until you have developed your own preferences.

JOIN THE CLUB
Pretty much the first thing I did, once I had bought the bike, was sign up to the Triumph Owners' Motorcycle Club, as it gives access to people with an unparalleled depth of knowledge about these old bikes. I immediately enquired by email about any history they might hold on my Bonneville, and got a reply from two club officials within a few hours, sending a scan of the colours for that year, and the date of manufacture: 13 April 1979. A quick check on the internet revealed this to be a Friday … good job I am not superstitious!

PROJECT MANAGEMENT
The restoration process will always be a juggling act between the conflicting forces of time, money and ability, with their eventual proportions differing with each individual restorer. All the skills needed to renovate a Triumph can be acquired by most people, but whether you have the time or inclination to develop them is a different matter.

Motorcycles can easily be broken down into large sub-assemblies, which can then be dealt with individually. Use a digital camera to record the process, as it will jog your memory later, when

INTRODUCTION

you struggle to remember where a cable ran, or if a washer had been fitted. It can be very easy to rush on, when tearing things apart, but keep stopping and taking those pics; your patience will be rewarded in the end. An old-fashioned notepad and pen is a good idea, as well, to list anything which needs replacing as you go along. Keep referring to the relevant Parts Book as you work; it will help to identify immediately any missing bits. Even better, print off the relevant page, and tape it to the bench next to you. A selection of freezer bags with a zip top, and an indelible marker are also indispensable, as you can keep bits from each section together. Buy lots of bags, and keep the contents of each down to a manageable amount.

PARTS

Finding bits for your bike will not be a problem. I typed 'T140' into a well-known auction site, and it turned up 13,000 listings. Finding good quality parts, though, will mean following recommendations from other owners and/or club members, which usually leads you to the door of one of the marque specialists. In general, it is best if you try and rescue as many original bits as you possibly can, as remanufactured items are often not up to original standard. People selling Triumph spares tend to use the factory parts numbering system, which takes a lot of the pain out of choosing the correct bit, and also allows some price checking to be done before laying out your hard-earned cash. This point should not be underestimated, as I found lots of bits on auction sites which actually cost more than buying direct from a specialist, sometimes by as much as 300%. I used LP Williams for most of my bits; its service was exemplary.

HAND TOOLS

There are only a few dedicated special tools that are required during the restoration. These are mentioned in the text, when their use is required. Most of them are relatively inexpensive, but a couple may make you wince. However, the damage that can be caused by *not* using them will cost considerably more. If you are a newcomer to British motorcycles, like myself, then you will need to populate your toolbox with AF tools.

0.9 Normally, I would never use cheap tools, but was persuaded to try these out. This entire ensemble, which performed faultlessly throughout the rebuild, cost less than half the price of a tyre.

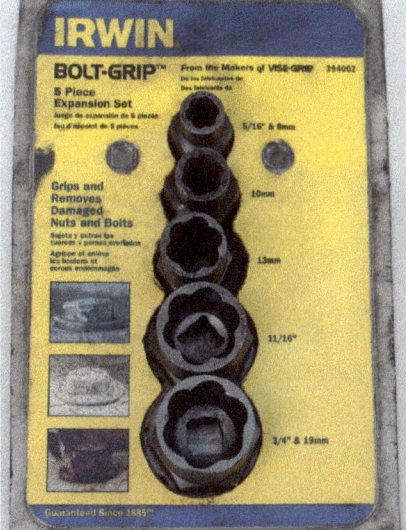

0.10 No matter which tools you arm yourself with, there may be some fittings that have been butchered previously, and nothing will sit on them properly any more. Time for a set of these, which will grip even the most rounded nut or bolt, and get them out.

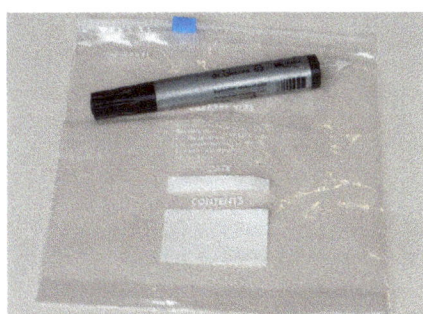

0.8 Clear food bags, with label fronts and a slide locking top, were ideal for keeping parts together, along with a suitable marker pen.

I would always advise choosing decent quality ones. It is much better to fork out on a good ratchet and a

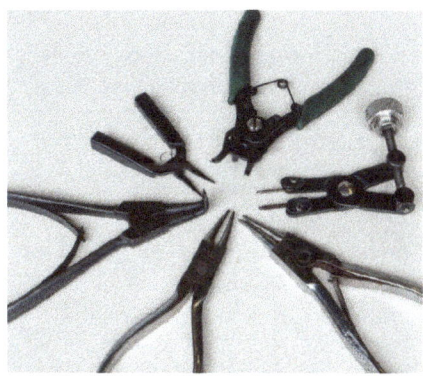

0.11 Any rebuild will almost certainly harbour a variety of circlips, so a collection of pliers to match would have to be sourced. Multi-function ones were cheaper, but inherently less sturdy. Locking ones were the most useful, but priciest. Right-angled versions were particularly useful on brake work.

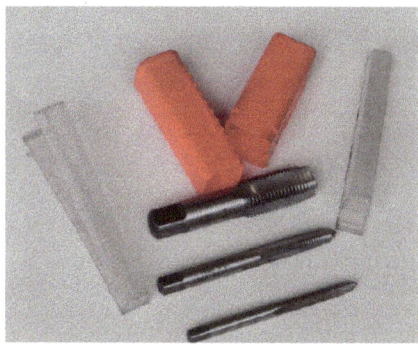

0.12 Taps are hopefully only needed occasionally during a rebuild. How often they are required, and the variety of sizes needed, dictate whether a set or individual items are more economic.

0.13 My toolbox was full of sockets, sadly all of them metric. It was time to lay out some cash on Imperial versions. This restoration lark can be expensive.

HOW TO RESTORE Triumph Bonneville T140

handful of quality sockets, rather than on a cheaper, giant special-offer set filled with sizes you will never use, and which will quickly break, having rounded off any tight fittings you might come across, taking the skin off your knuckles at the same time. Having said that, on the advice of a colleague, I was encouraged to try some very low budget sockets on this build, and, although their fit wasn't always perfect, they survived the restoration unscathed, and a rack of ⅜ sockets and allen key drivers cost less than £20. Chuck in something like a Clarke's Pro-range ratchet for similar money, and you have a very wallet-friendly but usable little kit.

BIG TOOLS

This is where the expenditure starts to creep up, unfortunately, so it depends on the depth of your pockets, and your desire to have the best. A hydraulic bike bench may be well worth the investment, as life is a lot easier, when you are not grovelling around on your knees struggling with recalcitrant fittings. The next investment might be an impact gun. If you want to use an air-powered one, or do your own paintwork, then a compressor is next on the list. If not, there are plenty of powerful battery versions. Both types make removing tight fittings a doddle, and they do it without imparting any great rotational forces in the process, which means locking bits up to undo them is no longer necessary. If you do buy a compressor, and can afford a big enough one, then who would *not* want a blast cabinet? A dedicated parts washer is preferable to a random collection of plastic buckets and an old brush, and an Ultrasonic cleaner is great for those gummed-up carbs. A plating kit might be a good idea, and a full set of polishing mops for your bench grinder. You do have a bench grinder, don't you? Oh well, better add one of those, too. Finally, a nice sturdy metal engineer's bench on which to rebuild that engine wouldn't go amiss. Tempted by that lot? If so, then you can say goodbye to another couple of grand of your hard-earned cash. Well worthwhile, if you intend restoring more than one bike; less so, if you lose interest once your pride and joy is resurrected.

0.14 An impact gun makes stripping components down much quicker and considerably easier. Once you have used one, you will never want to be without again. Battery versions are commonplace.

0.15 Air-driven guns are less cumbersome, especially the ⅜ gun on the right.

0.16 To drive them, though, you would need a compressor. Something like this is about as big as you're likely to want to go in a domestic setting, for both size and noise reasons. It will run air tools for short bursts, and handle paintwork tasks reasonably well.

0.17 A parts washer is a great restoration tool, as there will be lots of cleaning required. Filling one is a financially sobering experience, but cheap fluid would be very much a false economy, so bite the bullet and buy decent stuff.

MEASURING TOOLS

At the very least, you are going to need a torque wrench – and a general automotive one isn't going to cut the mustard, as bike fittings are mainly low range. Budget options are not much use, either, as their accuracy can be variable, which defeats the object. Digital calipers can be picked up cheaply, as can telescoping ones for bore measurement; neither will be particularly accurate at the budget end of the market, but will be enough to give you an idea of wear – once you think there is a problem, the part will have to handed over to an engineering shop for rectification, anyway, and it can do the really careful stuff. A decent quality multimeter should also be on your shopping list, for assessing that creaking old electrical system.

WORKSHOP SAFETY

Restoring any vehicle, especially a large and heavy motorcycle, will throw up potentially dangerous situations during the process. Do not rush. Take time to assess the task in hand, and visualise any potential hazards that may result. Read any literature supplied with power tools. Mains

INTRODUCTION

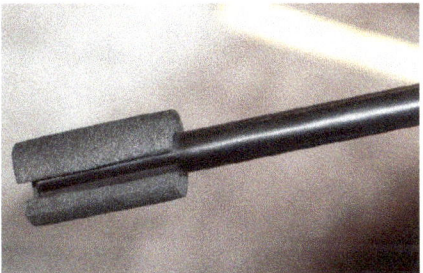

0.18 Think about potential hazards to your health, when working on your bike. Once up on a bench, I quickly decided that the bare handlebar ends were a danger, for example, so they were wrapped in foam sleeves.

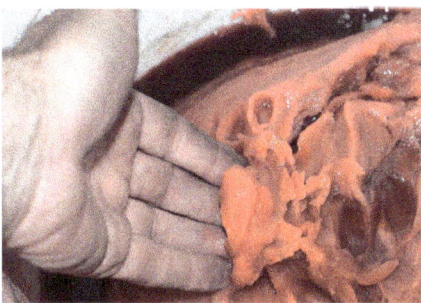

0.19 A decent hand cleaner is going to be needed. This one had micro beads to help shift the grime. Not the most environmentally friendly, but effective.

0.20 Drained fuel should be kept in a proper container before disposal.

0.21 You only get one pair of lungs, so if you intend painting, buy the best mask you can afford.

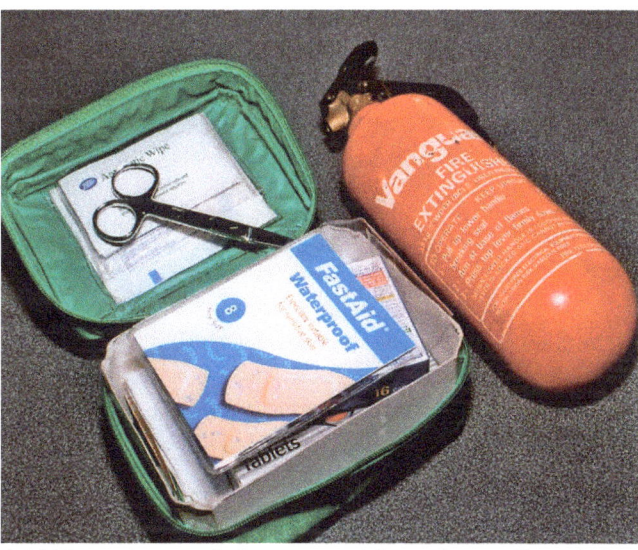

0.22 Two other workshop essentials: a decent first aid kit, and a fire extinguisher, both kept within easy reach.

electricity is a killer, so make sure that a circuit breaker is part of your set up. Fuel should be drained before work begins, and stored in sealed containers designed for the purpose, or disposed of completely. At least one fire extinguisher should be on hand at all times, as using heat to free-off seized components can be very useful. Make sure the workshop is free of flammable material, and always allow 30 minutes at the end of the day to make sure that nothing is quietly smouldering, just waiting to burst into flames once you have left. Protect your eyes and lungs from dust and fumes, and be aware that old brake linings could contain asbestos. Wear appropriate clothing for the task in hand: thick cotton is the best material for overalls; and wear stout boots. Use a barrier cream on your hands before starting, and a dedicated hand cleaner with a moisturising agent afterwards. Remember to protect the environment, as well: dispose of old oil, waste metal, plastics, etc, at your local recycling site.

DISCLAIMER

It is the responsibility of anyone undertaking work on their motorcycle to ensure that they are competent to do so; are aware of the risks, and have taken sensible precautions to protect their own health and safety. The author, publisher and retailer cannot accept any liability for personal injury, financial loss or mechanical damage as a result of any information included in, or omitted from, this volume.

THANKS

At times in any restoration, a helping hand is required to deal with a stubborn fixing, or to provide a second opinion as to whether the work is really up to scratch. My thanks must go to my mate, Tony, once again, for his assistance at virtually every stage of the process. Thanks, also, to the many Triumph owners that I collared in car parks and bike shows, who endured my many questions about the marque and their experiences with it.

Mark Paxton

Chapter 1
The plan

INITIAL INSPECTION

My Triumph didn't appear at all bad sitting in the autumn sun, but looks can be deceptive. The fuel was a stinking, sludgy brown slime, the carbs were stuck solid, and the petrol taps bunged up. The front master cylinder was empty of fluid, and the front caliper was solid. The rear brake was also seized, and, by the feel of the lever, so was the clutch. Cosmetically, it was let down by non-standard paintwork, which was peeling from what appeared, initially, to be a leaking tank, fitted with non-standard and partially missing decals. The frame had patches of scabby rust and flaking powder coat. The tyres were rock hard, although it was obvious that neither had seen any miles since being fitted, which made me very suspicious about why the bike had come off the road in the first place. There was no battery, just to round things off, but there were four packets of fuses, which indicated that electrical problems were sure to be on the list of defects.

RUN IT OR STRIP IT?

There are pros and cons to both approaches. It can be very tempting to want to hear your new project run, especially if it all looks to be in half-decent shape. If you cannot resist the urge, there are a few precautions to take before even attempting it. First, make sure the carbs are free to move on the throttle, and that there is no debris in the air filter boxes/intakes. Take out the plugs, put a couple of shots of oil, from a can, down the bores, and slowly push on the kickstart to check for movement. If the engine turns over, do a compression check. It is unlikely

1.1 Doesn't look too bad in the picture and very much the type of import that would be very attractive to restorers. Matching numbers, mostly original, and complete with low mileage on the clock.

THE PLAN

1.2 With the engine looking clean and in one piece, the temptation would be to fuel it up and let rip. I resisted, and was glad of that decision, later.

Breaking the rust bond will be easier with power tools, but, even then, releasing oil, ulitilising a heat source and, finally, using destructive tools, such as a hammer and chisel, or a grinder, might be required. When faced with a recalcitrant fitting, it may be better, to remove whatever it is attached to, and work on freeing it once on the bench. Stability must be a consideration, too, as a partially dismantled bike might suddenly fall, damaging you or expensive and hard-to-find parts. All fluids should be drained before working on the engine or braking system, but expect some residue to catch you by surprise, so be ready with paper towel or rags.

to be totally accurate, but would give an indication of anything seriously amiss. Hook up a slave battery, and, whilst the plugs are still out, check for a spark. If you get one, then the next item required is fresh fuel – probably best supplied from a separate dedicated reservoir, unless you know that the tank is clean and capable of holding fluid. If everything is working out so far, the last item before lift off is the most vital: make sure there is oil in the frame, and that the pump is returning it properly, in little spurts. This will involve lots of kicking, or putting the bike in gear, and pushing it around for a bit.

If all of that checks out, put the plugs back in, and give it a go. If it runs, listen for any unusual knocks or rattles and check for oil leaks. It isn't a good idea to leave it running for long, but hopefully you will have a better idea of the state of the engine pretty quickly. You could also check out the electrical components.

The negative side of trying to get it going is the possibility of doing some damage in the process. Plus, with any bike that has been sitting for some time, things like oil seals would have hardened, or picked up on their respective shafts when moved, after years of dormancy, so they would have to be changed during the restoration, come what may. On balance, I think just stripping it to see what lies within is the safest option, so that was precisely what I did.

GENERAL ISSUES
Unless you are very lucky, pulling a bike apart is going to throw up some common issues. The main one is corrosion, especially if the bike has lived in the wet of Northern Europe.

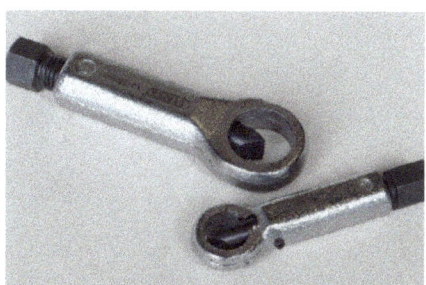

1.3 Nut splitters can be useful, too, but not worth spending lots of money on.

1.4 Light lubricating oil is a must. I normally buy it in large containers and use a hand pump, but I also check my local discount store as well, where aerosol versions regularly turned up.

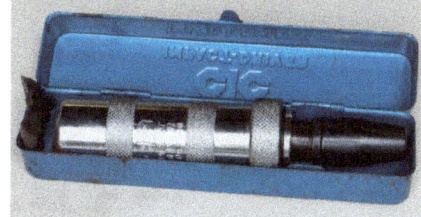

1.5 Impact drivers are pretty inexpensive, and an effective weapon against stuck fixings.

HOW TO RESTORE TRIUMPH BONNEVILLE T140

STRIP SEQUENCE

Before doing anything, I thoroughly degreased and washed the whole bike. It made the whole process easier, and reduced the risk of bits of crud ending up where they were least appreciated. The Factory Manual suggested leaving the bike in one piece to start off, so the rear brake could be used to hold things tight, when undoing some high torque fittings in the engine unit. Inevitably, this is not always possible – my bike didn't have a working back brake, but, if you have access to an impact gun, it doesn't matter, and even if you haven't, you can lock the crank through the conrod eyes, as we shall see.

1.6 Small items can be more frustrating than major components: this junction box for the throttle cables took a lot of wrestling to get it all broken down.

I removed the tank, seat and side-panels to gain a clear sight of what lay ahead. Stubborn fuel lines were cut, as they had to be replaced for safety anyway. I next removed the sticking calipers, but put the wheels back in, to allow the bike to be wheeled around more easily, in the short term. The mudguards and rear grab rail were taken off, too.

Carbs, exhausts and silencers came off next, to allow removal of the head. Control cables were unhooked, at the same time. The head and barrels were pulled to lighten the load. I was also going to take off the primary side, but changed my mind at the last minute, for no good reason.

The wiring loom and other electrical components were stripped off, to make sure nothing would interfere with the engine removal, as it was a very heavy lump.

The front wheel, forks and instrument binnacles came off next; then the handlebars and yokes. The rear wheel and shocks swiftly followed. The rear spindle can be a pain, unless you support the wheel as you try and pull it free. If it's sticky, put a block of wood under the wheel, and tap out the spindle using a soft drift.

Swinging arm, centre and side stands, plus various minor frame fittings, saw it all apart, with one final oddity showing up: there was a sumpguard fitted, which seemed odd for a road bike, but given the trapped mud between it and the frame, the last owner must have lived up a track, at the very least. It was clearly designed for the bike, and not homemade, but its origins were a mystery, until, by chance, one of the classic bike magazines ran an article on an Oil-in-Frame BSA Firebird from 1971, and there was the same sumpguard clearly visible. Maybe Triumph re-used it on its late off-road bikes, like the Tiger Trail, but I have yet to track down a decent photo of one.

Obviously, this was just one of many possible strip sequences, but it seemed logical, given the state of this particular bike. It was also pretty enjoyable, as the bike can be torn down fairly quickly, so the sense of achievement and progress was high. I tried to hang onto that feeling, as I knew that there would be times in any rebuild where things would not go quite as swimmingly.

The following chapters outline the actual disassembly and repair in more detail.

1.8 Take note of the routing of cables and wiring, as it might not always be obvious when it's time to refit. On many occasions, my memory failed me, so it was back through the hard drive to check the strip-down pictures.

1.7 The best laid plans often fall apart once the spanners start flying. I intended to strip the primary side, to lighten the load when it came time to remove the engine, but somehow it did not happen.

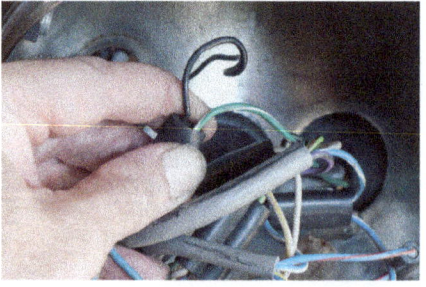

1.9 Take note of any discrepancies, as you go.

THE PLAN

1.10 Expect surprises along the way. Until the side panels were removed, I had not noticed that the air box had been junked and foam filters fitted by a previous owner.

1.11 Photos are especially helpful when it comes to non-standard items such as this oil filter installation, which would not be in any manual or parts book. I decided, pretty early on, that it would not be going back in such a prominent position.

1.12 This oil cooler was also non-standard, this time marked 'Ford of Canada.' I wasn't keen on all the pipework feeding it, so it, too, would not be featuring on the rebuilt bike.

1.13 Getting the bike to this stage was deceptively quick and easy. Getting the engine out was where the hard work began.

1.14 Properly dispose of all unwanted materials, especially oil. Your local facility will be able to recycle large amounts of your waste, if sorted properly beforehand.

1.15 All fluids must be drained before any major stripping, and stored in sealed containers.

1.16 Stuck studs or other fittings can try your patience: work through all possible options before reaching for a hammer.

1.17 Be aware of potential issues as you work, such as weight or balance: for example, will the engine tilt once the front and rear mounting plates are released?

Chapter 2
The frame

The Oil-in-Frame Triumphs, launched in 1971 got off to a rather shaky start. Immediate criticism of the seat height and a marginal oil capacity might have been overlooked by Meriden fans, had it not been for the fact that the frame had come from the drawing board of bitter rivals BSA. Within a year, it had lost a couple of inches from the seat height, by altering the rear rails, and pretty much remained like that until the end of production, although the area around the swinging arm and the arm itself got beefed up over the years.

Despite the initial moans, the P39 frame, as the factory designated it, was still regarded as superior to most of the competition throughout its life, and with good reason. It was hand-welded by skilled staff, on high precision jigs. On the road, it gave handling that was far superior to its Japanese rivals, and not far behind the best of Italian exotica.

PREPARATION
Once stripped of all mechanical parts, the bare frame needed to be thoroughly degreased. Mine, being relatively late, was originally powder coated from the factory, and the

2.1 With everything removed, the frame looked a lot worse than I had originally thought. There was also a considerable amount of congealed oil on the underside. Most importantly, though, it showed no signs of damage. The manual had dimensions, to aid checking whether anything looked out of line.

covering had cracked and peeled badly in some areas, allowing rust to get a grip. The big decision now was the best way to deal with it. Media blasting had been an obvious choice with previous bike restorations, but the thought of bits of grit getting into the oil-bearing frame was a little unsettling. If you decide to go down that road, ask the person doing the work how they intend sealing off all the frame apertures.

A second option was soda blasting. Considerably kinder to aged metal than hard media, it has the added benefit of leaving no solid residue to threaten the integrity of your mechanical parts. Unfortunately,

THE FRAME

2.2 A power washer is handy to clean the frame, or, if you have invested in a compressor, one of these would be a cheap, but effective cleaning tool.

2.3 Lots of solvent got rid of most of the old crud, but nooks and crannies needed scraping out with a screwdriver, before a second round of degreasing.

2.4 The original powder coating clearly demonstrated the drawback with that type of finish: noticeable flaking and subsequent corrosion.

2.5 The rust had spread a long way under the coating, out of sight.

relatively few places offer it, so your frame may have to travel some distance to be sorted out. Soda may also have trouble shifting the remains of old powder coat.

The third option was cheap and cheerful, but involves more manual effort. A wire brush on an angle grinder will get rid of old coatings of any type quite quickly. **Caution:** Make sure that your eyes are properly protected, and be prepared to pick stray wires out of your arms on a regular basis. In tight areas, abrasive papers will have to be used, which is time-consuming and awkward.

Paint stripper would work on earlier painted frames, but has a harder job shifting old powder coating, as modern formulations are a lot weaker than in days of yore. Buy it from a professional car paint supplier. It will almost certainly be stronger than branded stuff from a DIY superstore. The version I sourced was great, and lifted the old powder coat with ease, making the job a lot quicker than I had feared.

Things went pretty smoothly, but, as usual, by the time I reached the hand-sanding of the tighter sections, I was wishing I had just had it blasted.

2.6 De-rusting can be speeded up by using a knotted wire brush on an angle grinder, or perhaps a flap disc.

2.7 The wire brush was very effective, but snatched on fittings and shed a lot of wire. Eye protection and overalls were essential. It was slow going, so an alternative was needed.

2.8 I decided to use this unbranded stripper, designed for automotive, rather than domestic, use.

2.9 It ripped into the powder coat in a very satisfying manner.

2.10 The first scrape got rid of most of the original surface, unfortunately revealing yet more hidden corrosion.

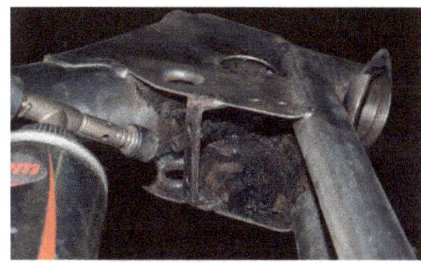

2.11 Some sections, such as the frame neck, were impossible to get into, so a blow torch was used to burn off the old coating, rather than rely on the stripper.

17

HOW TO RESTORE TRIUMPH BONNEVILLE T140

CRACKS AND DAMAGE
With the old covering off, the frame needs careful checking over: Tricky if you have left it with someone else for blasting and coating. The base of the oil-bearing down tube, where the swinging arm is mounted, is prone to cracking, and later frames received reinforcing, at that point, to try and stop it. The tube joints to the headstock should be looked at for signs of creasing or other damage, and the lower frame rail at the sidestand mounting for cracking or poor previous repairs. If aftermarket accessories are fitted, especially those employing U-bolts, check for frame crushing – there have even been reports of damage, where remote oil filters have been over-enthusiastically attached.

2.12 The base of the main oil-bearing tube was another ideal candidate for this method. This is the area to check for stress cracking. Mine had the later, reinforced frame, but I still had a good look over it.

2.13 The screw holding the old seat lock was a shear type, so a slot had to be cut in the top of the remains, which were then unscrewed.

POWDER COATING
If this was your choice, and there is a lot going for it, make sure the frame is internally as clean as you can get it. This will probably mean repeated rinsing out of the oil-carrying section with paraffin or degreaser, and some hard work with bottle brushes. The heat involved in the coating process may allow old oil left inside to ooze out and spoil the finish. Talk to local restorers to get up-to-date advice on the best person to entrust your frame to, as good preparation is essential. You cannot afford to just go for the lowest price. Ask to see samples of the finished colour, as the degree of gloss can vary tremendously. Discuss the options of light coverage at critical areas, like the swinging arm pivot points, and over the frame number, as heavy coats can cause problems. The other issue with powder coating comes where rust has left pock marks in the metal, as it can end up looking lumpy. Once you get your newly coated frame home, you will have to rinse it out a couple of times, once again, to make absolutely sure that it is harbouring nothing nasty.

PAINTING
This was the method outlined in the manual, and the one I followed, with a few additions. It has the added benefit that any pitting can be filled, to produce a perfect surface before painting, unlike powder coating The whole frame was cleaned with panel wipe, using one cloth to put it on and another to take it off. A couple of coats of wash primer (etch primer) were put on and allowed to

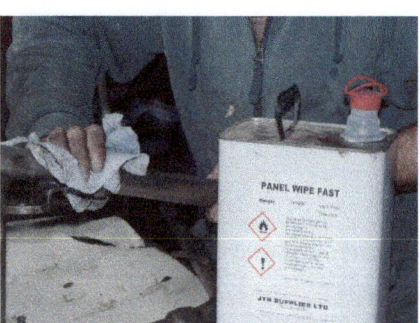

2.14 The last of the stripper was washed off, the frame dried, and then carefully panel wiped to make sure that it was perfectly clean and ready for painting.

2.15 Any pitting that still harboured rust could be treated with a suitable killer.

2.16 The oil filler hole and return pipe were capped, and then taped up. I also noticed that there was a small amount of final sanding to be done, which I had missed the first time round. This meant another round of panel wiping when I was finished.

2.17 The large oil filter hole needed capping, too. Luckily, a plastic coffee cup top was pretty much the right size, and needed only minor trimming to fit.

THE FRAME

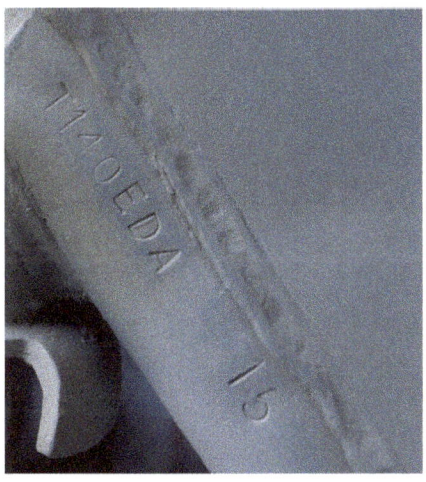

2.18 The entire frame was then treated to a coat of etch primer. If you need a dating certificate from the Owners' Club, now would be the time to photograph the number; it will never be as accessible or clear again.

go off for a couple of days. Once cured, the frame received several coats of primer, to build up some depth to allow flatting back to a smooth surface with 800s wet-and-dry paper used wet. The residue from the rubbing down was wiped off, and then it was all panel wiped once more. The top coat was in straight black gloss or, for an even higher shine, base coat and lacquer (as outlined in the section dealing with painting the tank later in the book) could have been used. If you do not have a compressor, all of these coatings, and many more, are available in aerosol form.

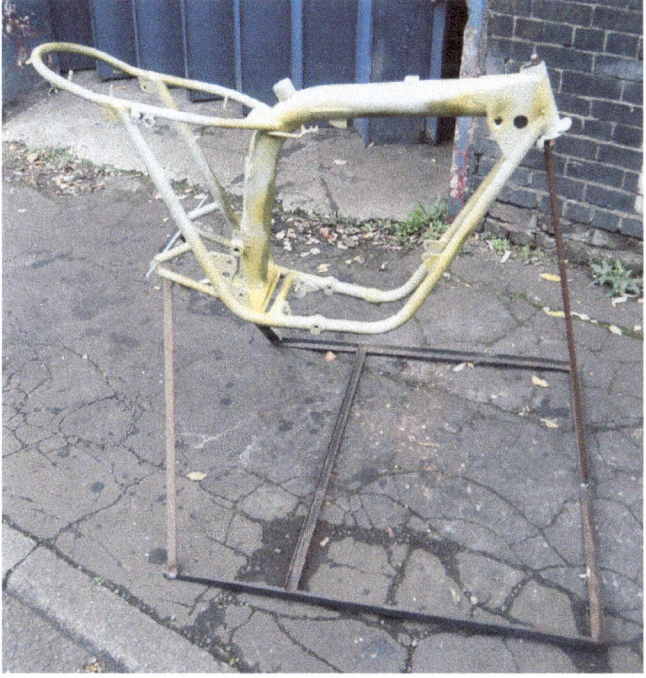

2.19 I fabricated a frame stand from a scrap road sign and some concrete reinforcing rod that had been abandoned nearby, which made it possible to get at all parts of the frame in one go. The first coat of primer was laid on.

2.21 The flatting was done initially with 600s wet-and-dry, well lubricated with soapy water. In the process, I got back to the etch coat in a few places, so, once washed off, the frame was primed and flatted once again, until I was happy it was all smooth.

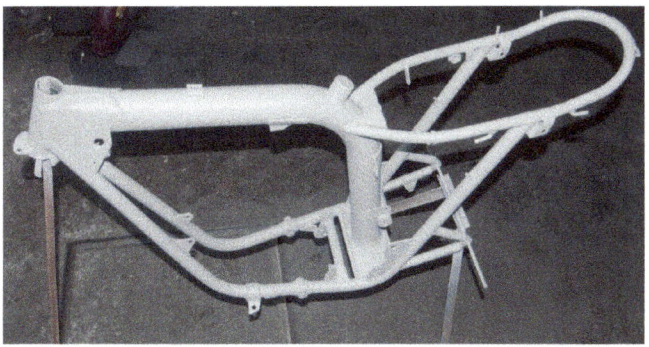

2.20 Several coats were applied to give a good depth, to allow flatting back to an even and smooth surface.

2.22 It then received several coats of gloss black, and ended up looking like this: perfectly presentable, not as rugged as powder coating, but a lot easier to touch up in the event of future scratches or other marking.

HOW TO RESTORE TRIUMPH BONNEVILLE T140

SWINGING ARM

The swinging arm was changed during production to accommodate the arrival of a rear disc, then the relocation of the caliper, and also a beefing up. It was checked whilst still on the frame, but, with the shocks removed, for side-to-side movement and smooth travel through its arc. Stripping and overhauling procedure remained the same as laid out in the manual. It is probably a good idea to remove the grease nipples before trying to remove the spindle, as it would not be the first time that the wrong length items have been screwed in place, causing binding. If the spindle is seized, a combination of heat and penetrating oil should be enough to get it moving. If not, the best solution is to have it pressed out, rather than end up using bigger and bigger lump hammers, especially given the frames' propensity for cracking in that area.

Bush removal and refitting can be done using a factory tool, or by a shouldered drift, according to the book, although the former (or a home-brewed version of it) is preferable. Replacements should be pre-sized, but I was warned by a couple of vendors that 'minor' reaming or honing may still be required. Getting the arm back in is a little fiddly, as you have to line up the bushes as you tap the pin through. It is also rather frustrating, if, like me, you manage to line everything up, and then remember that you haven't fitted the rubber sealing rings. Once done up to the correct torque, it was swung through its arc and pulled at 90 degrees to the line of the frame, just to double check before moving on.

2.24 Mine tapped out easily. If badly stuck, I would be tempted to have it pressed out, to prevent damage to the frame.

2.25 Cosmetically, the arm was suffering from exactly the same problems as the main frame, and was dealt with in the same manner. They can twist, but this one sat perfectly on a level floor.

2.26 There are pivot sleeves, which should simply push out. They can then be slid back over the spindle and rocked to give an idea of their state of wear.

2.23 The swinging arm was pretty conventional. I removed the grease nipples, top and bottom, before trying to remove the pin.

2.27 Once the sleeves were out, the bushes were exposed. These were in remarkably good condition.

2.28 On reassembly, the frame tube was heavily greased.

2.29 The only issue when tapping the pin back home was trying to keep everything lined up and in place, especially the four rubber seals.

2.30 Back together, the nipples were fitted with new fibre washers, and then the assembly pumped full of grease.

THE FRAME

OIL FILTER

At the bottom of the main tube on the underside of the frame lies the oil filter. It was under a finned alloy plate secured by four ½in nuts. It should be a metal gauze mesh, but many bikes have had the original unit swapped out for a paper element filter, as fitted to the BSA B50. There was some debate about the wisdom of this modification, as it relied on the oil pump having to overcome more resistance in the delivery path to the engine, which was why most people opt for an external unit fitted on the return side.

2.31 The oil filter unit sat at the bottom of the oil bearing spine of the frame under this plate.

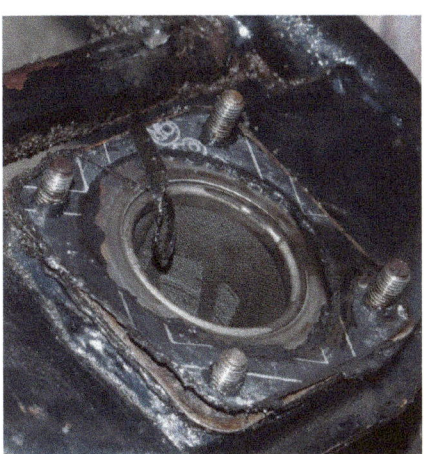

2.32 With the bottom plate removed, the filter unit was trapped between a pair of gaskets and, in this case, copious amounts of solidified silicone sealant – never a good sign.

2.33 The filter itself was in good condition, as new, in fact. They often end up split or squashed.

2.34 All the parts were thoroughly cleaned; there were new gaskets in the engine set I purchased, so it was ready to go back into the painted frame.

2.35 The mating surface on the frame was carefully cleaned and degreased. A thin smear of gasket sealant was then added, followed by the first gasket.

2.36 The side of the gasket facing the filter unit was then greased and the filter inserted. The bottom plate had its gasket sealed to the metal, but once again, I used grease where it met the filter unit, which would make future changes easier

2.37 The cleaned and completed unit, bolted back up to the frame, ready to go.

CENTRE AND SIDE STANDS

The centre stand was secured with two shouldered 9/16 bolts, which were held in place by a pair of 11/16 nuts, either self-locking like mine, or with tab washers on the earlier bikes. Removal was simple: I was able to hold the stand against spring pressure to get the bolts out. Once off, check that the bolt holes have not worn oval, or that the mounting lugs are bent. It is not unknown for the

2.38 The centre stand bolted on, although the design altered slightly during production.

HOW TO RESTORE TRIUMPH BONNEVILLE T140

2.39 The stand could be removed with the spring under pressure; as long as some caution was exercised, the securing bolts would still come out.

2.40 Refitting, though, was a very different story. The spring was so strong that, initially, I wondered if it would go back at all. With someone else holding the frame to steady it, the spring was pulled into place, but it was tough going.

2.41 The sidestand was easier, but once again, the spring took some effort to encourage it back into place. I didn't think this stand was original, and I really did not like its chromed finish. Fortunately, a black version was available.

stand itself to be twisted or fractured. Building it back up was altogether much harder, the manual makes it seem very simple, but the spring was seriously strong. Thick wire attached to a large screwdriver shaft, muscle, and an assistant to hold the frame were all needed.

BATTERY BOX AND COIL MOUNTING PLATE

The battery box was mounted on rubber bushes with a shouldered pin passing through them, the frame lug, and the coil mounting plate, secured by 7/16 nuts. Old age had caused the rubber to deteriorate on mine, and bond to the shaft of the pins, which made removal awkward, but, after a lot of spinning, the pins were persuaded to come out. The box was a tight fit when the mudguard is still on, and had to be wiggled out on the left-hand side.

2.42 The battery tray was suspended in the frame by three rubber bushes, secured by long shoulder bolts.

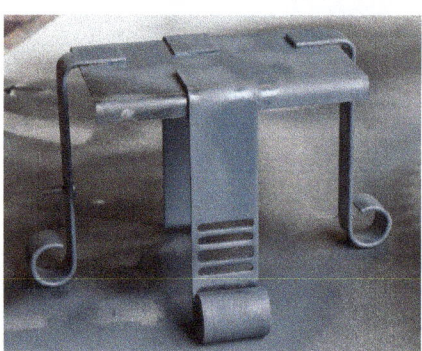

2.43 The tray was in a sound, but scabby, condition, so was etched and blacked ready for refitting.

2.44 The coil panel which sat next to the battery was also fine, but crusty, so it, too, was sanded and painted.

2.45 The bushes were lightly lubricated with rubber grease, and the unit bolted back into the frame.

HANDLEBARS

The handlebars were mounted in P-shaped eyebolts, which passed through metalastic bushes in the top yoke, and were secured with 9/16 nuts. The bolts also have rubber isolating mounts in washers; on reassembly, check they are assembled correctly against the diagram in the Parts Book, or they won't do their job. The ones on mine had been incorrectly fitted, which made the bars even more wobbly than they should have been.

There was a special tool for removing the metalastic bushes, which was a simple stepped drift. They can be very tight, though, and I chose to use the twin socket method initially – unfortunately, it didn't work. They were stuck solid, so an hydraulic press was called into action, which got them going with little effort; although, given their reluctance, I really think the drift method would either have failed, or done damage to the castings. The sleeves were quite slim, which made selection of an appropriate socket rather tricky. After painting, the bush seat was taken back to bare metal, greased lightly, and the new bushes pushed home by reversing the removal method. I thought that the vice would work on the way back in, when everything was clean, but I was mistaken; the new bushes were a very tight fit, so, to be safe, I used the hydraulic press once more.

A factory option, and still around, were solid mounts to replace the rubber ones, originally offered for use when a fairing was fitted, but they might also give a better feel

THE FRAME

2.46 The handlebars were rubber mounted, and unduly floppy on this bike, which would have to be investigated.

2.47 The bushes were worn out and had been assembled incorrectly. The order was laid out in the manual for reassembly.

2.48 The metalastic mounting bushes in the top yoke were also badly worn, and needed changing.

2.49 I initially tried drawing out the old ones using sockets and threaded bar, but to no avail.

2.50 I had to resort to using a hydraulic press, which did the job in seconds, but the choice of socket was tricky, as the edges of the bushes were very slim.

with the high and wide US bars, at the expense of numb fingers when running hard. Once all the rubber parts were replaced on mine, and the washers assembled correctly as per the diagram in the manual, everything ended up pretty taut.

STEERING HEAD BEARINGS

The Triumph head bearings were a modern design when the bikes were built: top quality and very long lasting. Unlike lots of bike restorations, they are not necessarily an automatic replacement.

With the wheel out, and the brake lines and cables removed, the yoke bolts were loosened – 9/16 allen keys on the top, and the same size bolts for the bottom, on mine. The handlebars came off; then the fork legs were dropped out. The top yoke

HOW TO RESTORE TRIUMPH BONNEVILLE T140

2.51 To remove the head bearings, the first job was to take off the plastic cap over the stem nut. On earlier models, this nut was larger, chromed and uncovered.

2.52 The allen-headed securing bolts, and the corresponding nuts securing the fork legs, were loosened, and the legs tapped down out of the yokes. They were a little reluctant to start with, so the slits in the yokes were spread in the traditional way, by tapping a screwdriver into them.

2.53 The pinch bolt securing the main adjusting nut was removed (arrowed), which allowed the nut to be unscrewed completely. The bottom yoke and stem were then withdrawn from the frame.

2.54 With the yokes off, the first thing underneath was a rubber seal, which lifted away ...

2.55 ... underneath, I was surprised to find what I thought was another rubber seal, but it was, in fact, part of the bearing, which was sealed.

2.56 The bearing simply lifted out ...

2.57 ... revealing the track, which was in perfect condition, apart from one small spot, where the weight of the bike had been resting unmoved for several years and marked it. Although not heavy, I decided that it was better to replace the bearings.

THE FRAME

2.58 The rollers themselves were fine, but still destined for the scrap bin, as they could not be paired with a new track.

2.61 The shouldered area on which the bearing sat was small, so it didn't take long to free it and slide it off the stem.

2.65 To get it on, a steel tube of the correct diameter could be used, or a slim punch, as in this case. This method worked perfectly well, as the interference fit was not particularly strong and, as long as you tapped evenly and only on the inner race, it would not harm the bearing in any way.

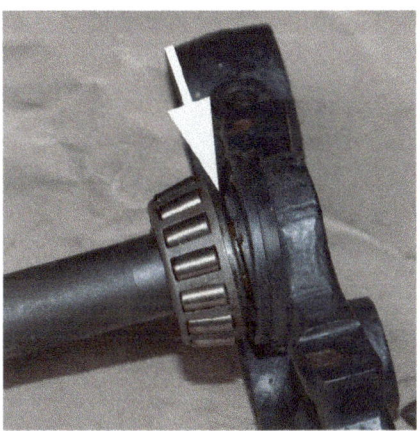

2.59 The old grease was cleaned off, and the bottom rubber seal peeled back, which revealed a narrow gap between bearing and stem …

2.62 With everything off, the bottom yoke was degreased and painted, ready for installation of the new bearing.

2.63 New bearings are not particularly cheap, but, well-greased, will last for years.

2.66 I made sure that the bearing was fully seated and square.

2.60 … which was how they would be removed. I inserted the blade of a thin chisel into the gap and gently tapped my way around the bearing, which lifted up the stem, until there was enough room to get the chisel onto it, side on.

2.64 The stem was greased, and the bearing and seal slid down until it was in contact with its raised seat, where it stopped.

2.67 The bearing was then greased, ready for insertion.

HOW TO RESTORE TRIUMPH BONNEVILLE T140

2.68 Before that, though, the tracks in the frame had to be changed.

2.69 They were recessed, and, to get them out, a long thin punch or similar drift was needed. The job was further complicated, as there was little to get the punch to sit on, thanks to a pair of abutment rings that were fitted under the tracks.

2.70 The only option was to drive both ring and track out together. Once started, they came out fairly easily, although the punch marks were clearly visible.

2.71 When I fitted the new tracks after paint, I made sure that the bearing seats were thoroughly cleaned back to bare metal, and put the new abutment rings in first.

2.72 To get the new tracks in, a dedicated bearing driver was a useful tool. If one isn't available, then the old track can be used to drive the new one home.

2.73 I made sure that the new track was completely square on to its recess, before applying any force to seat it.

2.74 It took several decent blows to get them into place. You could tell when they were fully home, as the hammer note clearly changed pitch.

THE FRAME

2.75 The top yoke was spread, whilst the nut was started, as it was tight and had a fine thread, and I did not want to risk damaging it.

2.76 The steering lock assembly was fitted at the factory using a shear bolt, which was then covered by a small cap. In theory, the cap can be removed and the remains of the bolt extracted. Mine was rusted solid, so had to be drilled out.

2.77 The threads were then re-tapped, to make sure they were clean and useable …

2.78 … although I shouldn't have bothered with all of that, as I was once again caught out by internet sales: the old lock, for which I had no key, is on the right; the supplied 'genuine' replacement, on the left, was noticeably fatter, and would not fit the yoke.

stem pinch bolt was undone and removed, and the plastic cap over the stem nut removed. Pre-'79 bikes will have a chrome split sleeved nut instead of the one in the pictures, but it does the same job, and is interchangeable if you prefer the look. The bottom yoke was supported whilst undoing the nut (1in spanner), as the stem would drop out of the frame, otherwise.

The tracks in the frame required a long drift through the neck to tap them out. Straightforward enough, but the edges were slim, as they were partially shrouded by spacer (abutment) rings, so both had to be knocked out together. When replaced, the areas where the tracks sit were made perfectly clean, then greased lightly. The new track was left in the freezer for a couple of hours, then tapped into place using a bearing tool (or the old bearing track as a drift) having remembered to drop the abutment rings in first. The new tracks were carefully squared up, before trying to tap them in any distance, or they would have jammed and needed excessive force to keep them moving.

The top bearing race just drops in, but the lower has to go down the bottom column. It drops easily until nearly home, but the last little bit requires some gentle persuasion. A length of suitably sized tube would do the job, but I did not have any to hand in the workshop. The alternative was to use a narrow flat-nosed drift, and gently tap the inner bearing track down. It did not require much effort, I just made sure that it went down evenly.

Once the bottom yoke is back in, the top one had to be spread slightly to start the stem nut, as it was tight and I did not want to risk stripping the threads.

STEERING HEAD LOCK
Like most of these locks on '70s bikes, as a security measure, it was little more than an ornament. To replace it, there was a small cap which has to be drilled through, and underneath, the remains of a shear-headed grub screw. Mine was reluctant to come out, even using a dedicated extractor, so had to be drilled out completely. Once gone, the lock was knocked out of the yoke. The screw threads for the locating screw were then cleaned up using a ¼in by 26tpi cycle thread tap, ready for the new shear bolt. The replacement lock was lined up, but it immediately became apparent that it was the wrong one, as it was a couple of millimetres too big – another auction site purchase, I should have just bought it from a specialist. Unwilling to hold up the build, I left it for the time being: it will be replaced later, when I find the correct one.

Chapter 3
Forks and shocks

FORKS

Like the frame, the front forks were partially a product of the experiences gained by BSA during its successful off-road antics, combined with a nod to the Italian manufacturers of the time, and a gentle progression from Triumph's back catalogue. Supple, with decent travel, they were certainly well up to the task, with weak seals and a degree of stiction being the only real defects.

Removal

Their attachment was conventional, and the manual sets out a removal procedure which I chose to ignore in many respects. The road wheel was secured by alloy caps on four studs at the bottom of each leg. These caps are prone to damage if over tightened, so I examined them closely, to make sure that they were fit for re-use. The top caps on my bike's legs were screw-in plastic (plated steel in earlier models), and one was missing. Under them were alloy plugs with a $7/16$in allen key recess built in. These can be tight, so I decided to crack them open, whilst the legs were securely held in the yokes. The pinch bolts, top

3.1 The front forks didn't look at all bad, but their action was peculiar. I put this down to the seals having hardened or swollen from sitting for so many years. A strip down and check was essential.

3.2 The mudguard was removed (six bolts), then the lower fork caps undone to free the front wheel.

3.3 These caps were carefully checked, as the light alloy easily cracks, if they have been overtightened in the past.

FORKS AND SHOCKS

3.4 The fork top caps were unscrewed. These were plastic on my model. Previous models had chromed metal versions.

3.5 Underneath each cap lurked a screwed plug with a hex recess for removal. This held the fork internals in place. For better access, the handlebars were removed.

3.6 The plugs were just cracked open – but left in their threads, in case I had to tap the legs down and out of the yokes, later. I was also concerned that I might not be able to get enough purchase on them, once the legs were on the bench.

3.7 The top yoke fork clamp was a hex-headed bolt.

3.8 The lower had a nut and bolt, which passed through a securing tag for the fork shrouds.

3.9 With the fork legs tapped down (which was easier on one leg than the other), the shrouds could be removed and the headlamp ears unbolted for painting later.

3.10 With the forks out, the strip down could commence.

and bottom, were removed, and the first leg was knocked down and out with a hammer handle. The other, inevitably, wasn't going to give in that easily, so a screwdriver was gently tapped into the split in the yoke, to open it slightly, and copious amounts of releasing fluid sprayed around the yoke eyes.

Stripping and rebuilding

Once on the bench, the dust seal was removed. Then the $7/32$in allen-headed bolt at the bottom of the leg was undone, using an impact gun (this overcame the need to use a special holding tool). Spring tension may be enough to allow it to be undone manually. If not, the top plug will have to be removed and a tool rigged up to stop the internals turning as you undo the bolt. This will require a $13/16$ socket and a long extension bar to reach down inside the leg.

With the allen bolt out, the top plug was removed. The oil, which was then tipped out, was a chewing-gum pink, so clearly a modern formulation, rather than the originally specified ATF. The spring was removed; then the stanchion was pulled out from the lower fork leg. The damper rod, which was reluctant to move, had to be tapped out gently. Check for the Dowty washer – mine came out with the rod, but it often remains lodged in the lower leg, if so, it will usually come out when the leg is degreased.

The old fork seal was levered out using a dedicated tool, which, thanks to its shape, rested on the rubber at all times during the removal. If you use something else, such as a large screwdriver, make absolutely sure that it does not sit on the alloy, or it may dent or even crack it, so use a steel spacer to spread the load.

Everything was then degreased and cleaned before inspection. My stanchions were great, with only minor discolouration on a tiny part of the chrome. If they had been pitted, new ones are inexpensive, although I personally would choose the slightly costlier versions made in the UK or Italy, over those originating in the Far East. Another option would be to have the originals hard chromed – even pricier, but with almost certainly the best finish long-term.

HOW TO RESTORE TRIUMPH BONNEVILLE T140

The forks were an un-bushed design, so wear in the lower leg would have been terminal. The clearance between leg and stanchion should only be a couple of thou, so a visual inspection for scoring, and a wiggle test to spot excessive wear was needed. Alternatively, the measuring tools shown in the engine chapter could come in handy, once again.

The new seals were the 'leakproof' type, which are held in place by metal rings. Reassembly, as they say in all the best manuals, was the reverse of disassembly, taking care that the inner lip of the seal was greased and the stanchion inserted with a slow twisting motion. The bottom bolt was tightened, with the damper rod nut held by the socket on a long extension, mentioned previously. The spring was dropped in, and 190cc of 10W synthetic fork oil added, before the top cap was screwed in place. A set of Norton Commando fork gaiters were fitted, as the UK climate is not kind to chrome. The ones I bought turned out to be made from very thin rubber, so fitting was easy, although I wish I had bought the genuine Norton ones, which were around 50% more expensive, as I don't hold out much hope of the current ones lasting very long.

Refitting
Getting the forks back in was pretty much the reverse of getting them out, remembering to fit the chrome shrouds between the yokes, first. The leg that was tight coming out remained just as stubborn in the opposite direction, despite both yoke eyes being carefully cleaned and greased. The slits in both upper and lower yokes had to be opened up, and the bottom of the leg given a few firm taps with a soft mallet, but not before the bottom fork studs had their nuts put back on flush, to make sure that there could be no possible damage.

3.14 The top plug removed alongside the bottom bolt. The fork oil was carefully tipped out of the top before any further stripping down.

3.15 With the dust seal out of the way, the stanchion was ready to be pulled out of the lower leg. The fork seal fitted did not look the correct type for the year of the bike, but it didn't matter – my replacement wasn't going to be either.

3.12 To strip the forks, this hex-headed bolt had to come out to free the stanchions. If you have an impact gun, it will rattle them out easily; if doing it by hand, the whole assembly inside the legs might turn and refuse to allow the bolt out. The solution is pictured later.

3.16 The rod pulled out of the leg – but without the Dowty washer.

3.11 The fork oil could have been drained at the very start of the removal process, but I was concerned about the state of the drain plugs, which were rusty and are always prone to chewing up, so I left them until the legs were on the bench, for better access.

3.13 Either way, the top caps had to come out. This one was a bit chewed up from previous removals, but still came out without drama.

3.17 It looked like this, and was removed by inverting the fork bottom, and tapping it on the worktop.

FORKS AND SHOCKS

3.18 The valve assembly and spring were then removed. At first glance, everything looked original and in good condition.

3.19 There was a seal in the middle of the assembly; this has been blamed by many for the 'stiction' these forks are reputed to suffer from. I suspect that modern seals may be manufactured from superior material, so the issue may not be as prevalent now …

3.20 … it had to be changed anyway, so the old one was removed used a sharp pointed pick.

3.21 As a replacement, I went for this fibre anti-stiction alternative from LP Williams. More expensive than a simple rubber seal, they were recommended to me as an obvious upgrade whilst the forks were apart.

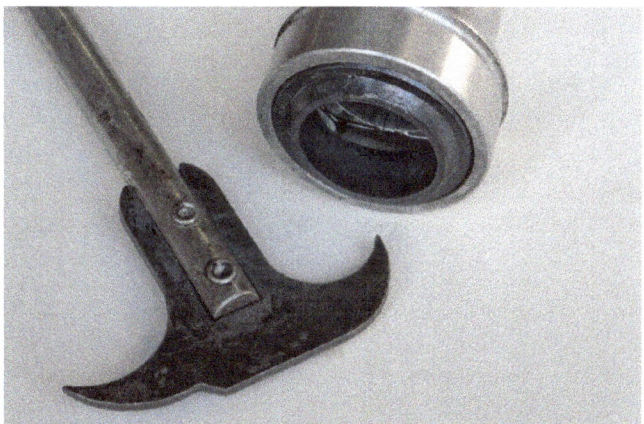

3.22 Oil seals can be very tight, and even more so when they have lain undisturbed for more than a decade, so use of the correct tool is highly recommended.

3.23 Its shape means that, once hooked under the old seal, the point of leverage remains on the rubber, not the light alloy of the fork leg, which could burr or crack, rendering it useless.

31

HOW TO RESTORE TRIUMPH BONNEVILLE T140

3.24 Even with the correct tool, the seals were really tough to shift, but came out eventually.

3.25 With the old seal out, it was obvious that the last change had been carried out using a screwdriver or similar, which had gouged the alloy. Unfortunately, the seals had been refitted over this, without any attempt to address the damage first, which would not have helped their longevity. I spent 15 minutes with some abrasive paper, and removed the high spots.

3.26 The stanchions were carefully checked: all I found was a very small area with some minor discolouration.

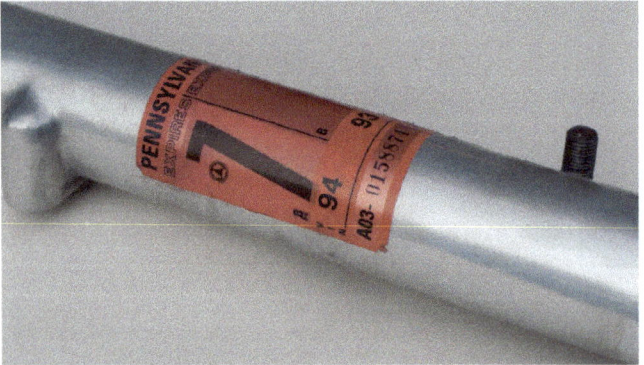

3.27 The fork bottoms needed cleaning and polishing, but first, this sticker had to be removed. I assumed it would peel off, but no chance.

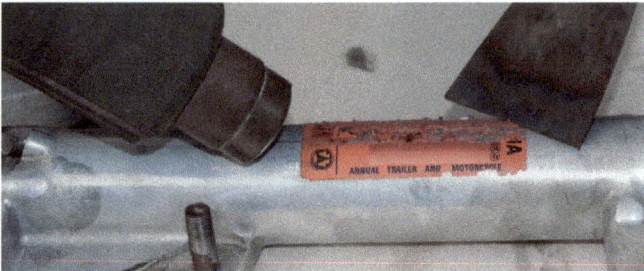

3.28 I resorted to a hot air gun, and a scraper, lightly applied to prevent any chance of gouging the leg. This method heated the alloy up very quickly, so I had to wear gloves.

3.29 For polishing, I used a small kit bought at an autojumble, mounted on an old bench grinder. The stall speed was low, so only light pressure could be applied; patience was the keyword.

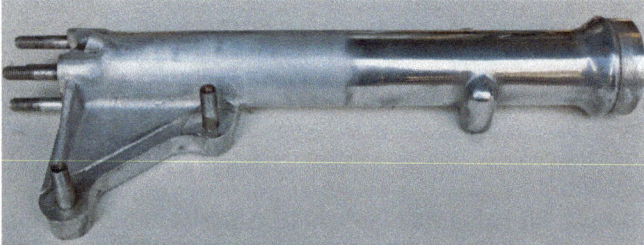

3.30 This DIY setup can produce perfectly good results, though, as this partially done leg shows.

3.31 I wondered how good a shine could be achieved, so I kept going to find out. The reflection was pretty clear for a piece of polished alloy. It was far too good, though, for my desired look, so I cut the shine back a bit afterwards, with some slightly abrasive polish.

FORKS AND SHOCKS

3.32 With the leg polished, it was time to start the reassembly. First up were some new seals. I opted for 'leakproof' types. It would appear that these are loved or derided in pretty much equal measure by owners I spoke to, but I thought I would give them a try.

3.34 The securing ring does not need to be driven fully home, as this type of seal needs a small amount of space to move up and down to be fully effective.

3.35 The fork spring was measured and came well within specification. They can sag over time and miles, and, if replacements had been needed, a set of Progressive Springs were repeatedly suggested as superior to the original set up.

3.36 The stanchion was carefully inserted through the new seal, which was lightly oiled to ease progress. A slow twisting motion on the tube as it initially went through the seal helped enormously as well. The innards followed, with the Dowty washer stuck to the bottom of the rod with grease to make sure it located properly in its recess in the lower leg.

3.33 I fitted the seal and its locating ring with a dedicated seal driver. If there isn't one in your toolbox, then a suitably sized socket could be used as a drift.

3.37 The hex-headed bolt on the bottom now had to be screwed in, to hold everything together, but, as mentioned previously, the damper rod would spin unless this nut was held firm …

HOW TO RESTORE TRIUMPH BONNEVILLE T140

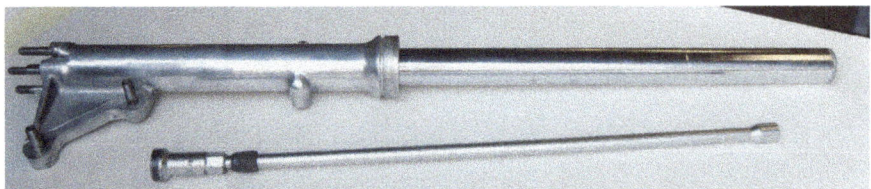

3.38 ... there is a special tool to do that job, but a long extension bar and socket is just as good. This bar had a locking collar to hold on the socket; if there isn't one, the leg could be held upside down to keep everything in place whilst you work.

3.39 The rusty drain screws were replaced by new items, along with their alloy washers.

3.40 The forks would have originally been filled with ATF in the '70s, but I decided to go with a modern synthetic formulation. 10W is about the correct weight to match the original specifications, but obviously could be adjusted to suit your chosen riding style and weight.

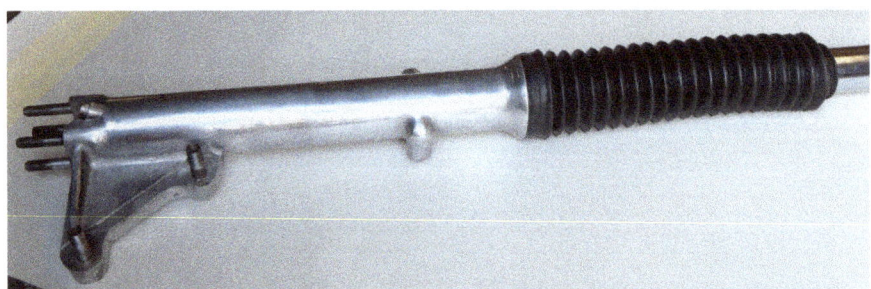

3.41 I opted for fork gaiters, as the chrome needed as much protection as possible here in the UK. Not original, but I actually quite liked the look.

3.42 New plastic fork tops were purchased, along with their stickers, to finish the job.

3.43 The new caps were not easy to get started on their threads, and were then tight all the way down. At least that would help prevent them vibrating off later.

REAR SUSPENSION UNITS

Five different shocks were fitted to the T140 during its production run, and are listed below. If originality is not an issue, then the choice of replacement is obviously huge.

Standard Girling

These were fitted from the start of production until the 1979 model year, and were mounted in traditional fashion. They cannot be overhauled, so restoration tasks are limited to cleaning, repainting, and fitting new mounting bushes. Aftermarket versions, which have the correct look but not necessarily the quality of the originals, are available at very reasonable prices.

Gas Girling

These followed on from the earlier shocks, but are immediately identifiable by their 'upside-down' orientation. They will only work this way up. Once again, they are not rebuildable.

British-made Hagon shocks are available with the correct rod-down orientation, are gas-filled like the originals, and can be tailored at the factory to suit your weight/use requirements.

Marzocchi Strada

Traditional in style, these shocks were very popular as an aftermarket fitment for a wide range of bikes back in the day, with many Japanese roadsters sporting them as upgrades. They benefit from being rebuildable, which is handy, as they seemed to need doing regularly on my bikes back then, but it may not be worthwhile, unless absolute originality is required. An overhaul kit can be hard to track down: look for one marked-up for 10mm rod shocks – usually expensive – but if you can find an industrial seal shop, you can probably get something matched up for considerably less. If the mounting rubbers need changing, only pricey remanufactured ones are available. Throw in some damping oil, and a replacement set starts to look like better value. The final issues are the chrome on the damping rod. This is thin, and does little to prevent rust getting a grip, which is terminal. If they have been rebuilt before, the

FORKS AND SHOCKS

32mm nut which seals the body is shallow – and appears to be made of cheese, so has probably been rounded off previously. I have seen several with new flats filed into them to compensate, which is effective, but unsightly.

Marzocchi Strada AG

These were fitted to relatively few models, and can be easily identified by the piggy back remote reservoir, filled with air under low pressure, separated from the oil by a rubber seal. These were top quality shocks in their day, and still command high prices, even secondhand, as many Italian bike owners are after them for their restorations. There are few modern remote reservoir options that look anything like these, so rebuilding may be the only answer, if you want the original look. Seal kits are still around, but can be expensive. There is also still the problem of weak chrome on the damper rods.

Paioli

These were only fitted to the very-rare TSX model, and later to the Harris-built models. Simple shocks, like the basic Stradas mentioned above, but with the same parts and damper rod issues. Probably a match up job for O-rings and seals, with items such as original top collars and mounting rubbers being almost impossible to find.

Shock rebuild

The shocks fitted to my bike were in good condition, apart from some paint loss, and rotting rubber mounting bushes, so a strip and cosmetic tart up was all that was required. They are held on by 9/16 bolts, which should be shorter at the bottom mount. The springs were removed with a special tool, which was one of such feeble quality that it only just managed to do the job, before retiring hurt to a dusty corner of the workshop. Very poor indeed.

The shock bodies were rubbed down, primed and painted, with the only complication being the adjuster sleeves, which were seized. After much twisting and gentle tapping, they eventually relinquished their grip, and were slid up over the damper rod to get access to their inside face for

3.44 My bike, like the majority of T140 production, was fitted with Girling shocks. These were highly regarded at the time, and often used as an upgrade by riders of Oriental machines.

3.45 To safely remove the top collars on the shock absorbers, a special compressing tool was needed. This one was purchased on the internet.

HOW TO RESTORE TRIUMPH BONNEVILLE T140

3.46 The end bent on its first outing. Will I ever learn? Answers on a postcard …

3.47 My shocks appeared to be in good condition overall, but the mounting rubbers had perished over the years.

3.48 The easiest method to remove parts like this is to use threaded bar and a couple of sockets. Tighten the nuts, and the smaller socket pushes the bush through into the larger one.

3.49 It was a really quick operation. However, as can be seen, I had to change the smaller socket halfway through removal, as the original would not push fully home.

3.50 Once out, the bushes and their sleeves were in even worse condition than I had initially thought.

3.51 Replacements were not too expensive, although their finish was not that impressive.

3.52 The shock absorber bodies were sanded and etch primed, ready for paint …

FORKS AND SHOCKS

rubbing down. They remained a tight fit, which made getting them back down without scratching the new paint a very slow and careful exercise, despite greasing them internally first. The new bushes were lightly lubricated with silicone grease on the leading edges, before being pressed back into the eyelets. This was done in a vice, as I decided that this method was less of a risk to the new paint. Once they were back together, they looked remarkably smart, and fit for many more miles.

3.53 ... which was applied with them suspended on cord, to allow complete access to all sides.

3.54 The new bushes were then carefully pushed into place, using a socket and a vice. The springs were simply cleaned before refitting, as the chrome was perfectly acceptable.

Chapter 4
Brakes

Given the age of these motorcycles, a complete overhaul of the braking system would be a very good idea, unless the bike came with documentary evidence of recent work. **Caution:** Brake pads and shoes may contain asbestos, so make sure everything receives a good soaking in brake cleaner to damp down any possible dust before and during disassembly, then carefully bag the old parts before disposal. Spilled brake fluid may damage paint and plastic, so remove the tank first and cover the clocks, then take extreme care when removing components to avoid splashes. If the bleed nipple is free and the hydraulics working, attach a rubber tube to it, and pump as much fluid out as possible, before starting any disassembly. Finally, the brakes are obviously a safety critical part of the rebuild, and the work is relatively straightforward, but do not attempt to overhaul them, if you feel in any way uncertain of your abilities. Brand new master cylinder assemblies, which are complete and ready to bolt in, are out there, as are the cylinders on their own, some in stainless steel.

The calipers and master cylinder assemblies are essentially much the same front and rear, so disassembly and repair are covered together. Some parts of the rebuild do not strictly adhere to the Workshop Manual recommendations; the reasons for that are laid out, and the final decision on the best procedure rests with each individual.

MASTER CYLINDERS
Up to 1979, the front master cylinder was moulded into the front half of the switch unit wrapped around the bars. It was removed by undoing four crosshead screws to split the switchgear, although the screws are often corroded, so some releasing oil or an impact driver may be needed. After '79, the cylinder was

4.1 The front master cylinder will either have a pipe which fits directly into it, or a right-angled union under a rubber boot, like this one.

4.2 The assembly was held onto the switchgear by hex-headed bolts on my model.

4.3 Union seals are copper. In theory, they can be annealed and reused, but check for signs of ridging due to over-tightening. If it is heavy, or you are in any doubt at all, just change them, which was what I did.

BRAKES

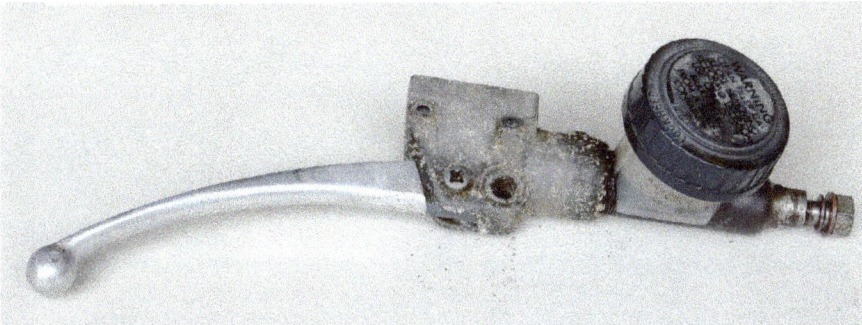

4.4 With the assembly off the bike, its cosmetic condition was poor – to go along with its lack of function.

4.5 The rear master actually looked better, but was still not working. Removal was done by the book, but was awkward and slow.

4.6 Both the steel cylinder and the alloy housing were corroded. Before going any further, I carefully counted the number of exposed threads: this would make reassembly a lot easier.

4.7 Inside the reservoir was a small nut and washer. Once released, the transparent plastic body could be lifted away from the cylinder.

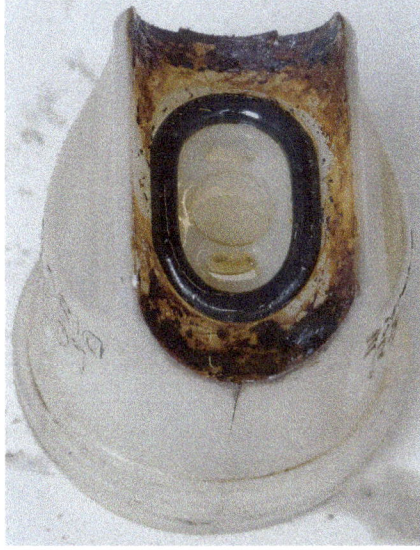

4.8 On the underside, there was a seal which would have to replaced later. It can be removed from its groove with a slim-pointed pick.

4.9 The rear caliper had a saddle, secured by a large nut. This had a lock washer tab which needed folding down to allow the nut to be undone.

on a separate bolt-on plate, secured by socket-headed bolts, which were far easier to deal with, despite suffering from corrosion like their predecessors.

According to the Workshop Manual, the rear master cylinder from 1977 was not user serviceable and, if defective, must be replaced. My 1979 model, though, was fitted with a cylinder which could be stripped, (but given the remains of a paint marking on the mounting plate, it may well have been a replacement from a breakers), so I went ahead and did so. Separating the master cylinder from the alloy housing was a major problem, requiring heat, and the old trick of tightening it slightly using the flats machined on the end (⅝, but a sloppy fit, on mine – a 15mm gave a better grip), before trying to undo it. Before removing them, count the number of exposed threads, and also count the number of turns needed to unwind them, as it will ensure accurate and quick reassembly. The internals of both were removed in the same manner, and, once apart, a large quantity of congealed black slime oozed out: neither had done much braking for a very long time. After a good soaking in degreaser, and a thorough clean, the bores were actually in very good condition and definitely reusable.

Reassembly was straightforward, and by-the-book. The only issue was getting the circlip in place, as it meant compressing the piston against spring pressure, which is a three-handed job, made worse by the need to use a screwdriver through the circlip you are trying to compress and fit, so the end of the unit was lightly held in a vice whilst I did it.

The rebuilt cylinders were wound back on to their respective mounts, and the grub screws reinserted, with a little locking compound to make sure they stayed there. If you are using new cylinders, and therefore have no witness marks/thread count to rely on, the procedure for setting the correct position is laid out in the manual. Some aftermarket cylinders are supplied with their own instructions, as they differ from the factory method. Either way, it sounds daunting in print, but, with the unit in your hand, it is pretty logical.

HOW TO RESTORE TRIUMPH BONNEVILLE T140

4.10 Under the saddle, there was also a seal, just like the front cylinder reservoir shown earlier …

4.11 … the securing nut assembly included a further seal. All of these were automatic replacements during the rebuild.

4.12 The next job was to remove the lever, so that the two sections could be unscrewed. The reservoir was still on, in this picture, as the actual strip sequence was not critical to this point.

4.13 On the underside, there were tiny grub screws through the alloy, to trap the steel cylinder in position. The alloy on this one was damaged, but not enough to stop the screw having a solid grip.

4.14 The alloy body was held in a vice; after the application of copious amounts of releasing oil, a spanner was used to unscrew the steel cylinder. This worked perfectly on the rear, but the front required some heat from a gas torch, applied to the alloy, before it finally moved. I counted the number of turns needed to unscrew the cylinder, as it came off, as a double-check (along with the exposed thread count done earlier.)

4.15 Once off, the first thing visible was a rubber cap which simply peeled off.

4.16 Under that was a piston, held in place by a circlip. The eyes were very small, but, using a suitably sized set of pliers, it was removed.

4.17 Once that was out, the spring pressure should have pushed the internals out …

4.18 … the piston came out easily; the remainder of the assembly needed a lot more knocking on the bench, before it was finally persuaded to show itself.

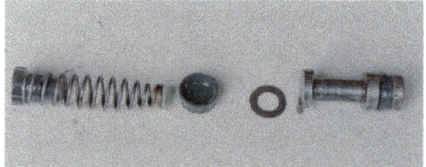

4.19 This was what the whole lot looked like. It was complete, and did not appear to be damaged in any way.

4.20 Both cylinders came apart in the same manner, and both were in surprisingly good condition internally. The exposed areas at the top had suffered from corrosion, but the working area, where the seals ran, looked fine. This was confirmed after a gentle clean-up with some fine wet-and-dry paper, well-lubricated with releasing oil. Definitely rebuildable.

BRAKES

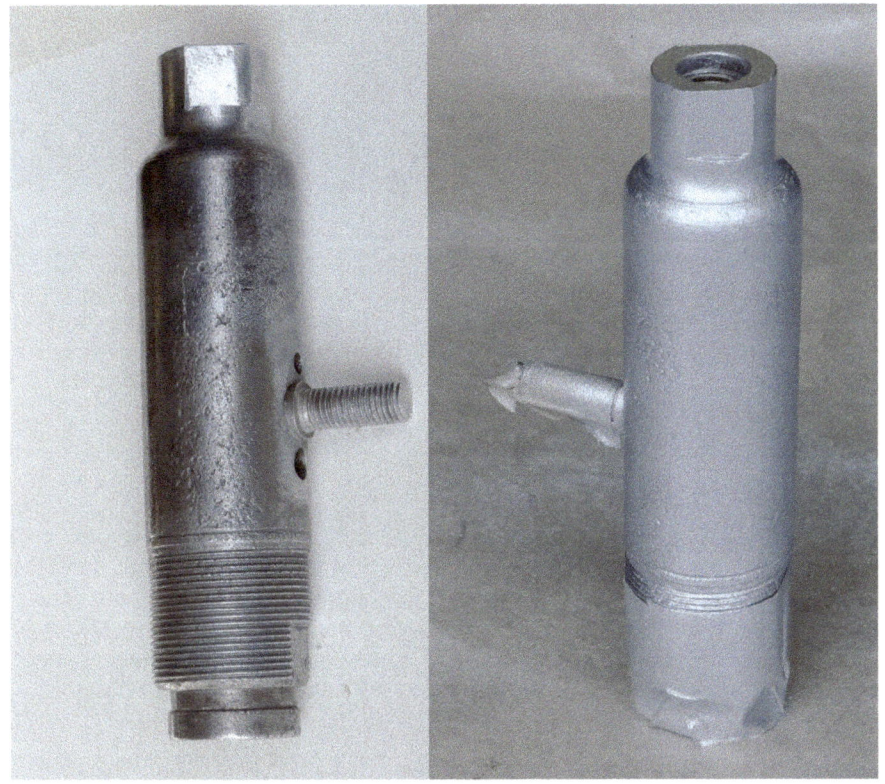

4.21 Once stripped and cleaned, the steel cylinders were de-rusted, using a soft wire brush on a bench grinder; then coated with silver wheel paint, which is a tough and long-lasting coating.

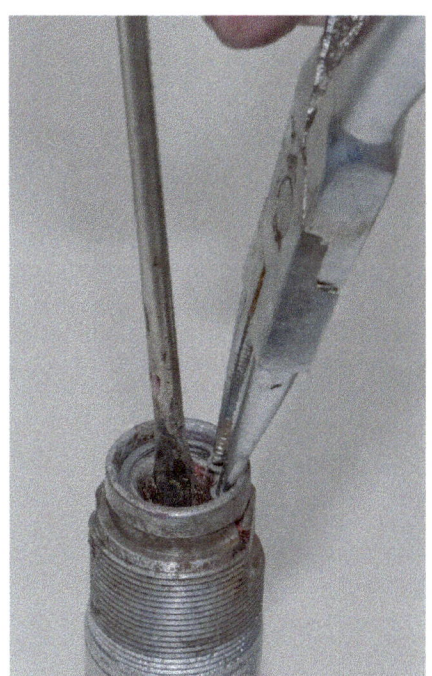

4.24 The piston had to be inserted against spring pressure, then held down, whilst the circlip was compressed and inserted into its slot. This was slightly tricky, and is probably best done with the body held in a vice.

4.22 Seal kits are cheap and pretty comprehensive.

4.25 When I refitted the reservoir, I made sure not to forget the small metal sleeve – it can easily be overlooked.

4.23 The new components were fitted to the old piston, making sure they were the right way round. They were then lubricated with red rubber grease ready to go back in. The pitting on the piston was not a concern, as it was lighter than the contrast of the digital picture suggests.

4.26 The new seal was round and the seat oval, so it took a minute to roll it correctly into position.

41

HOW TO RESTORE TRIUMPH BONNEVILLE T140

4.27 The metal pushrod of the front cylinder was stuck in the alloy housing ...

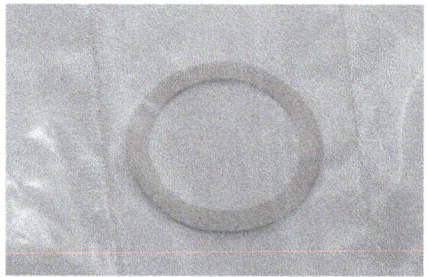

4.31 The only thing missing from the rebuild kit was this paper seal, which fitted under the reservoir cap.

4.33 This is even more essential on early bikes. These have a black fluid reservoir, which requires the cap to be removed to check the level.

4.28 ... it took gentle persuasion with a slim screwdriver and lots of lubrication to get it out.

4.32 It may not seem important, but it stops the rubber diaphragm sticking to the cap when it was unscrewed, and so cuts down the risk of spillage.

4.34 The rear cylinder housing also needed cleaning, and the adjusting nuts freeing off and lubricating, before refitting.

4.29 Lots of elbow grease was needed to clean up the alloy, but both it and the lever were deemed suitable for reuse, by the end of it.

4.35 The seal was removed before the top-securing nut assembly could be de-rusted.

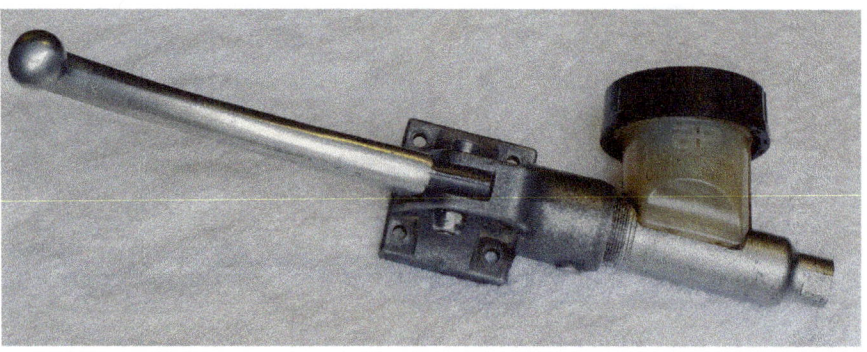

4.30 The cylinder was screwed back to its original position and, with it all back together, it was ready for many more miles. If cosmetics had been a critical issue, then complete, brand new assemblies are available.

4.36 The saddle seal was replaced, and once again, just like the top reservoir seal, it needed a little persuasion before it sat squarely in its recess.

BRAKES

CALIPERS

As with the master cylinders, if you do not wish to overhaul your calipers, then remanufactured Lockheed units are on the market, as are bolt on replacements from Grimeca. The Workshop Manual and period Lockheed literature both stress that the caliper should not be split, but, having looked at the units, I could not really see any reason why not; they appear completely conventional. The second issue was that only brake fluid was recommended for cleaning, which may have been sufficient when the bikes were not very old, but something stronger would definitely be needed today, so brake cleaner was definitely on my shopping list. Front calipers were held by 9/16 nuts on studs; the rear was the same size, but with nuts and bolts through the mounting plate.

Compressed air was used on both calipers, which got one piston moving in each. Both pistons were then reinserted, and held in place with self-locking pliers. This allowed all the effort to be directed to the other piston, which came free with a resounding pop, so make sure your fingers are well out of the way. Replacement pistons are readily available, as are stainless steel versions, although I was warned that some of the latter could be a frustratingly tight fit.

The seals supplied in the rebuild kit were symmetrical, while the manual refers to tapered versions and therefore to their orientation, so check and follow the appropriate instructions depending on the type you were supplied with. The dust seal retaining rings were the only troublesome part of the rebuild, as the new ones refused to fit on the rear and having, struggled to get them to seat properly, I decided to try one of the old ones instead. Fortunately, I had removed them carefully, when the calipers had been stripped, and it immediately slotted into place perfectly. Strangely, the new rings fitted with no problems on the front. The caliper half-securing bolts had their threads cleaned to remove old locking compound, then were reinserted with medium-strength thread lock, and tightened to 32lb-ft.

4.37 Fully assembled and bolted to its mounting plate, the rear brake assembly was ready to fit.

4.38 The only tricky part, when mounting it, was fitting the brake lever to the pivot. It was a tight fit, and tended to push the whole assembly through the frame rather than slipping over the pivot pin.

4.40 Mine had come cable-tied to the frame, which obviously was not correct. I assumed that it should be mounted to the air box, which was missing on my bike, so I bent up a small metal bracket, which, once painted, did not look out of place and was secure.

4.39 The rear reservoir was remotely mounted. It, too, had a seal between the plastic and the alloy body. If the pipe needs changing – mine was damaged – the replacement must be made of the correct material to be resistant to brake fluid.

4.41 The finishing touch was a set of new decals for the caps, which brightened them nicely.

43

HOW TO RESTORE TRIUMPH BONNEVILLE T140

4.42 The front caliper was covered by this chromed plate. A peculiar styling feature to my eyes, as I am used to seeing calipers in all their naked glory. It was fitted to most bikes, but a few left the factory without it, apparently.

4.44 Brake pipe unions are best removed using a dedicated spanner. The ones on this bike were not especially tight or corroded, but for something that has spent its life in damper climes, it might be a different matter.

4.45 Bleed nipples are especially prone to corrosion, but, once again, I struck lucky and they came out easily. If sheared, the only option would be to drill out the remains, which can be tricky.

4.46 Both calipers on my bike were seized. This one in the rather odd position of having one piston in, and the other out.

4.43 On my bike, the rear caliper was underslung. On later models, it moved to an upright position.

4.47 With the pads out of the way, the outer metal dust rings can be seen around the edge of the piston. They were carefully levered out.

BRAKES

4.48 Before any further attempt at removal was made, the pistons and surrounding area were cleaned, so progress could be easily monitored.

4.52 If you are not going to split them: rather than try and grab the pistons, take out the one in the opposite half to the bleed nipple, first; then seal the pipe hole with another nipple or bolt.

4.49 I used compressed air to get the pistons moving. If air had not been available, a grease nipple could have been welded onto a brake nut, and grease pumped into the caliper to force out the pistons. This is a very old method, but it works perfectly.

4.53 With them both tight, apply air through the arrowed hole and the second piston will be forced out.

4.50 Usual piston movement wasn't equal, so the loosest had to be clamped to divert pressure to the other to get it moving. Some jiggling around between the pair was needed, before they both started to move freely.

4.54 Once out, the old fluid looked disgusting, although the pistons were surprisingly good; not the rusty mess I had expected.

4.51 There wasn't room for both to come out together, but I was going to split the calipers anyway, so it did not matter.

4.55 The dust seal (which I had left in on this caliper) and the main seal were now accessible for removal using a thin pick.

HOW TO RESTORE TRIUMPH BONNEVILLE T140

4.56 The seal grooves looked in decent shape, as well. Had they not been, the caliper might not have been rebuildable.

4.60 The calipers had been split, as mentioned previously and, between the halves, was a seal which must always be renewed.

4.64 The piston, too, was lubricated. Overkill perhaps, but better than a seal turning, and having to strip it all down again, later. Excess grease would just dissolve in the brake fluid.

4.57 A quick wipe with a cloth showed the pistons to be reusable, too. Careful use of metal polish brought back an unmarked surface.

4.61 A new seal kit comes with all the necessary components.

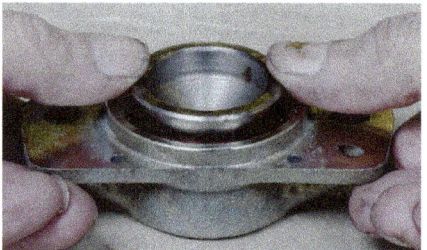
4.65 The pistons had to be inserted carefully, to make sure that they were square before any pressure was applied to seat them, or the seals would be at risk.

4.58 The old murky remains of the original brake fluid were removed in the degreasing tank.

4.62 The grooves and seals were lubricated with red rubber brake grease.

4.66 Once nearly home, the dust seal and metal securing ring could be fitted. Seating the ring proved a little tricky. The new caliper half seal had been fitted and lubricated at this point.

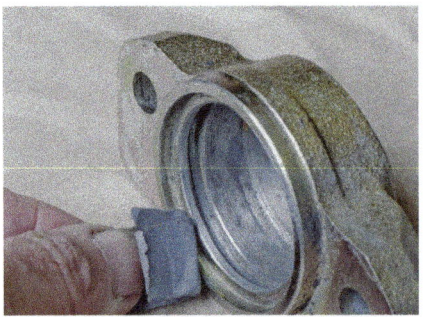
4.59 The surfaces and grooves were carefully cleaned using wet-and-dry, lubricated with brake cleaner.

4.63 Once in place, more was added, to ensure the pistons would slide in with as little resistance as possible.

4.67 The two halves, ready to go back together: I left the original anodised yellow finish untouched, as it was not in bad condition, and matching it with paint can be tricky.

BRAKES

4.68 All nipples and pipe nuts were treated with copper grease to help prevent corrosion. They were carefully tightened, just enough to seal correctly: over-tightening could causes problems later in the bike's life.

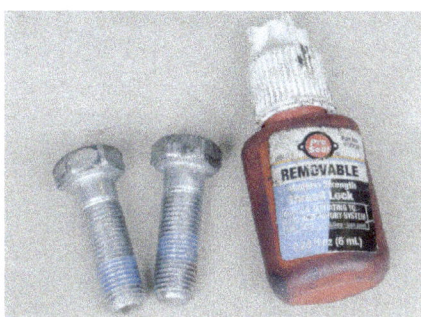

4.69 The caliper-half securing bolts were cleaned and refitted with some thread lock.

4.70 The overhauled unit, ready to go back onto the bike.

4.71 The chrome cover had a sticker in the centre, which was damaged ...

4.72 ... it was peeled off and a new one fitted.

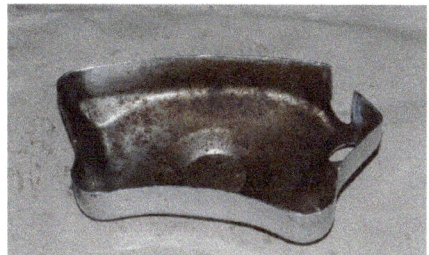

4.73 The chrome on the cover was in good condition, but the back was starting to rust. A chemical killer and zinc-rich paint would hold it down, or perhaps brushing with one of the wax-based rust inhibitors, for use on the underside and cavities on classic cars, would be a good idea.

4.74 Back in place, and fitted with new pads, the newly overhauled brakes should be as good as when the bike left the factory.

4.75 The rear caliper was not in such good cosmetic condition, so the old surface was removed using a wire brush on a bench grinder.

4.76 Wheel silver is not strictly an accurate finish, but would be hard-wearing.

4.77 Painted and fitted with new hoses, and bolted on to its repainted mounting plate, the rear caliper was ready to go ...

4.78 ... and, once bolted back into place, doesn't look bad at all.

47

DISCS

The discs fitted to these models were originally chromed, which helped keep them looking bright, and was also the market norm at the time. Unfortunately, it did little to improve their performance once subjected to rain, so alternatives may be better on a regularly used bike. Cast iron units make sense, and are the most common now, but, if appearance is paramount, then hard-chrome versions, similar to the original type, are also out there. My discs appeared to be in good condition (other than chrome loss), with little scoring and still a good thickness, so my first choice was to have them skimmed. Once bolted up in a lathe at the local engineering shop, it quickly became apparent that they were also warped, so the scrap bin beckoned. I chose UK-made replicas of the originals, from LP Williams as its seemed to be well finished compared to others I had looked at. Lightened and drilled versions are also an option, but I decided on the original look. When

4.79 The front disc's main issue appeared to be the loss of its chrome surface: sadly, that turned out to be the least of its problems.

4.80 The rear was more heavily marked and rusty, but I still held out hope for rescuing it. I was sadly disappointed.

4.81 New discs can come protected by some sort of grease, wax or oil film to prevent corrosion in storage. This has to be removed with brake cleaner or other suitable solvent, such as panel wipe. For maximum effect, wipe on with one cloth and off with another.

4.82 Original discs had their centres and rims painted black, so I decided to follow suit. I did not mask them, but simply wiped the overspray off with a cloth soaked in thinners.

the new ones arrived, I checked them against the old, and discovered a size anomaly. It appeared that the original 10in versions were changed, around 1980, to a smaller metric size, losing 4mm or so in the process, and my replacements were obviously the later size. Pad manufacturers sometimes only list one for all years, but apparently they can be fitted to either disc without any real detriment.

Models with alloy wheels were fitted with different discs. The T140D, running Lester wheels, has a six-hole fitting, whilst those with Morris alloys, eg Royal wedding, TSX, TSS, and optional on other models towards the end of production, have five-hole fittings. Both types are currently remanufactured, despite the relatively small numbers involved.

BRAKE PADS

Fitting the pads was straightforward, although it is easier (on a full rebuild) if you leave the fork spindle caps loose until the pads are in, to allow maximum wiggle room. Ferodo pads seem to be the ones most recommended, but, whichever type you choose, pad rattle may be an issue. If you take the spring from a retractable biro and slide it over each locating pin, apparently it sorts that out cheaply and efficiently.

REAR DRUMS

Relatively short-lived, the conical rear drum was conventional, with shoe replacement fully covered in the Factory Manual.

4.83 The first models came with this type of rear drum. Shoe replacement is covered in the manual and is completely conventional.

BRAKES

BRAKE HOSES AND PIPES

Old brake hoses will show their age externally through surface cracking, but time will also have reduced their strength internally, which will either allow them to balloon under pressure, or block up reducing flow. Lockheed recommended replacement every three years/40,000 miles. Mine were all in poor shape, so it was decision time once more. Original-style hoses are cheap, or there is the option to upgrade to a flexible stainless steel braided line, which would be superior in every respect, but does not have a sufficiently classic look for many owners, myself included. A black version is available, which would have been less obvious, but I had not seen one 'in the metal' at that point. The copper sealing washers used at joints can be annealed and re-used, but they really should be checked for dishing or ridging from previous over-enthusiastic tightening, which was precisely what had happened on my rear master cylinder. Where the flexible hoses pass through brackets, they should be secured with a serrated washer and a 9/16 nut. Metal pipes are 7/16 at the male end and 9/16 for the female.

4.85 When fitting the new hoses, I made sure that the anti-vibration washers were in place, to stop the securing nuts shaking loose.

4.86 Those securing nuts should be well tightened, but I didn't overdo it with the brake line union nuts.

4.84 Brake hoses should always be changed on any restoration, in my opinion. However, if you think they may not be old on your project, try doubling them over: this will reveal cracking, although it cannot show internal deterioration.

BRAKE LIGHT SWITCHES

The hydraulic inline brake switch unit at the front on my '79 model was held in place with 5/16 bolts, through a thin bracket which needed to be supported when undoing the unions, and, again, on reassembly. The switch unit itself can just be unscrewed from the main union block for replacement (1in flats). Previous models had the switch built in to the right-hand handlebar assembly, and was prone to breakages, as the operating pin was thin plastic, and the soldered wire joints also broke.

4.87 The hydraulic brake light switch simply unscrewed from the union. Its flimsy mounting needed support whilst that was done, and whilst the pipes were removed and refitted.

49

HOW TO RESTORE TRIUMPH BONNEVILLE T140

4.88 The rear brake light switch screwed into the frame, so replacement was easy – which was no bad thing, as I was warned that they often fail.

BRAKE FLUID
Debate rages about the best fluid for your classic. I don't intend to add to it here, as the choice is down to the individual owner. I opted to use DOT 4 as it was cost effective, able to meet the relatively light braking requirements of a classic bike (or at least one ridden by me), and it was on the shelf at my local parts shop. It will have to be changed every 18 months, as Lockheed suggested, to minimise any potential water damage.

BRAKE BLEEDING
I was warned that the rear brake circuit, in particular, would be a major pain to bleed, thanks to its underslung position, so I naturally approached the job with some trepidation. I filled the reservoir with fluid, leaving the top off afterwards, so I could see what was going on. I made sure the surrounding area was covered with an old towel, to contain splashes. I then operated the brake pedal fully, several times, very slowly. This was rewarded by the expected stream of bubbles, as the fluid displaced the air in the lines, which also confirmed that the cylinder had been assembled correctly. After a minute or so of this, I opened the bleed nipple, but only by a small amount: too much, and air can creep back through the gap in the threads. I lightly covered the nipple with my finger and pushed the pedal down, which released fluid. I then pressed down hard on the nipple, before I released the pedal; then repeated this several times. The nipple was then tightened, with the pedal down. A couple of pumps saw the beginnings of a brake, so a few more and, again with the pedal held down, the nipple was opened, then retightened, once pressure had been released, but before allowing the pedal back up. Another couple of times like this saw an operational brake, so the job turned out be quick and drama free. On the front, the same system was used and, despite it being a bit of stretch between lever and caliper, it too was bled very quickly.

4.89 The best choice of brake fluid would be the one that matched the needs of the bike, my riding style, and the depth of my pocket: DOT 4 suited me fine.

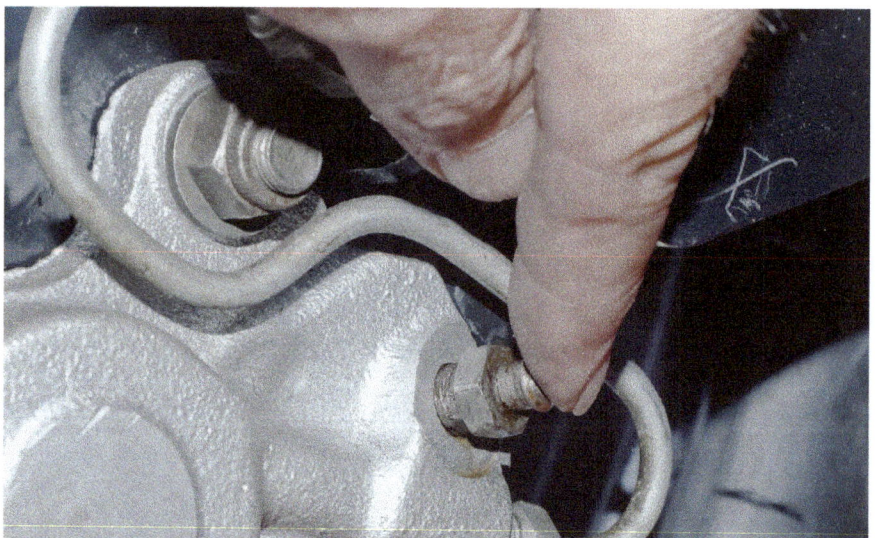

4.90 There are lots of brake bleeding devices on the market, but I have always relied on this very cheap and immediately available one, and have never failed to get pressure.

4.91 After the brakes had been bled, I applied pressure to the circuit, held it on, and closely examined each joint for signs of leakage.

4.92 When the caps were refitted, I made sure that the diaphragm was correctly seated, as per the manual.

Chapter 5
Engine and gearbox

The factory Workshop Manual covered the engine teardown and rebuild pretty well, but the passage of time would have made some of those procedures rather more tricky. The pictures and captions follow the strip down, so I have limited the text to areas where difficulties may be found, and possible solutions.

TOP END STRIP
The first issue raised its head immediately, and it was apparently a common one. The nut in the head steady captive plate was very rusty, so lots of lube, with the bolt worked repeatedly in and out, saw it eventually free. The cylinder head bolts were undone, in the same order as the tightening sequence laid out in the manual, a bit at a time.

The barrels, though, proved a bigger problem. I had bought a base nut spanner ready to tackle the 12-point nuts, but unfortunately, after removing two, the effort proved too much for it, and the centre of the spanner started to lose its edges. A lot of work with a small file, and a brief mental request to the patron saint of old British bikes were enough to get the last two off, before the spanner end was fit for no further action. I spun it over to remove the plain nuts and the ring on that end snapped at the first attempt. Unbelievable – and a salutary warning about buying unseen from the internet.

With the securing nuts out of the way, the washers were carefully prised from position, and the barrels

5.1 With the exhausts and carbs already stripped off, it was time to start on the engine.

HOW TO RESTORE TRIUMPH BONNEVILLE T140

5.2 The tappet covers can come off; mine were held by socket-headed bolts.

5.3 The rocker boxes were held by nuts on the underside, which were undone first …

5.4 … with the larger through-bolts left until the end. This sequence minimises strain. The other option would have been to back off the valve adjusters before releasing the rocker box fixings.

given an initial tug. Nothing doing, so lots of aerosol lubricant and smart blows square on to the side of the fins, and I was still rewarded with zero movement. An overnight soak, followed by more tapping, saw it free the next morning. Had that not been successful, as there were no obvious leverage points where damage could be avoided, I would have tried fitting four of the plain base nuts back on, with another nut jammed above it under the fins, and wound the lower one up to put it all under light tension, before applying heat to the cylinder flange at the base, especially around the stud holes.

There was a special tool to pull the gudgeon pin out of the pistons, or you could make one yourself from threaded bar, a couple of sockets and washers. The quick way (and the method I used) was to gently warm the pistons with a hot air gun and tap the pin out, but I made sure the rod was fully supported at all times, to prevent any strain on it or the big ends. I used a small punch to identify each piston, once they were removed ready for cleaning.

5.5 Under the boxes were a couple of sleeved head bolts (black) …

ENGINE AND GEARBOX

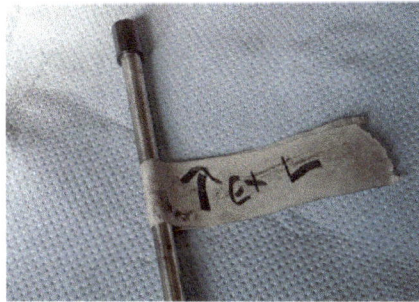

5.6 … before undoing them, though, I decided to remove and tag each pushrod in turn.

5.10 The fit of the tool was good, despite its low price.

5.7 With the pushrods out of the way, the head bolts could come out.

5.8 A wiggle revealed that three of the four sleeved bolts were loose – not good.

5.11 With the bolts out, the head was given a couple of gentle taps with a soft hammer, side-on to the finning to ensure no damage. This freed it without problem.

5.9 There is a tool sold for the bolts, although I had a suitable allen key socket already.

5.12 The head was lifted, but would not clear the studs, as the pushrod tubes were stuck to it. They took some persuasion to release their grip, but, once free, the head came away …

53

HOW TO RESTORE TRIUMPH BONNEVILLE T140

5.13 ... which revealed the pistons, that were carbon coated, but not too thickly, and free of any obvious damage.

5.14 The cylinder base nuts in the centre of the barrel were 12-pointed, and should be removed with a special spanner. I bought this one from a web-based supplier.

5.15 It fitted, and I managed to get the first couple off ...

5.16 ... although, by the third fitting, it was obvious something was wrong: the centre had started to round off. The other end was used for the outer nuts and, once again, something didn't feel right: the ring had snapped. At least the outer nuts were more accessible, so a conventional spanner got them off.

5.17 The tappets were loose in their guide blocks, and would fall out when the barrels were lifted off the case.

5.18 A piece of rubber (in this case part of the old, split gearchange boot) was jammed between the tappets to hold them in place.

ENGINE AND GEARBOX

5.19 I would like to say "then up it came," but the reality was lots of tapping and squirting of lubricant, combined with large amounts of muttered threats. The time gap between this and the previous step was around 12 hours.

5.20 Before the barrels were fully lifted, I inserted some packing material to stop the conrods dropping onto the cases. If I hadn't been doing a full strip, I would've used rags to make sure no bits of broken ring could drop into the engine.

5.21 The pistons showed signs of considerable blowby, but it was light so obviously it had not been happening for long. (The contrast of the digital image makes it look worse than it actually was.)

5.22 Whilst in the vicinity, I checked how level the case halves were, as there was some scope for misalignment if they had been apart before, but these were fine.

5.23 The circlips were prised from their recesses and the gudgeon pins tapped out, having supported the rods throughout, to prevent strain on the bottom end. The pins were lightly marked but not grooved. I labelled the pistons to prevent mix-ups. This temporary marking was later replaced by a couple of light punch marks.

TOP END INSPECTION/ OVERHAUL
Threads and studs

As each component was removed during the strip, the state of the threads was carefully assessed.

5.24 A set of chasers is useful to check out each thread as the disassembly continues. I noted any problems for rectification before reassembly.

HOW TO RESTORE TRIUMPH BONNEVILLE T140

Kits are available for home repair of any that are sub-standard, but are relatively expensive, especially if only a few sizes are needed, so it may be cheaper to get a local engineering shop to sort out defects. Studs would have to be replaced with new, unless the threads could be rescued with a chaser. High stress ones, such as the main cylinder studs, must be in perfect condition.

Rocker boxes

As I stripped the rocker boxes and cross checked with the Parts Book, I spotted another problem: the washers were arranged differently. Checking the manual clouded the issue, as it was different from the Parts Book, and like the set up on my bike. The internet came up with the theory that the order was officially changed in 1969, but Meriden ignored it and continued assembling the bikes incorrectly for another decade or more. If that was indeed the case (and factory service bulletins appear to back it up), then the relative position cannot be too critical, or the factory would surely have had feedback on the resulting failures and altered its assembly procedure; a decade is a long time to leave things, if they were causing damage.

5.26 Its neighbour, though, was a problem: the end was badly worn and pocketed, with the case hardening having given up the ghost.

5.27 The rocker arm washer assembly also did not match the diagram I had to hand, so some further investigation was needed.

5.28 The rocker shaft had to be driven out of the head from the oil-feed end.

5.29 To protect the threads, and prevent anything getting into the hollow shaft, the dome nuts were screwed back on.

5.30 A few taps with a soft hammer and the shaft started to come out, revealing the oil seal.

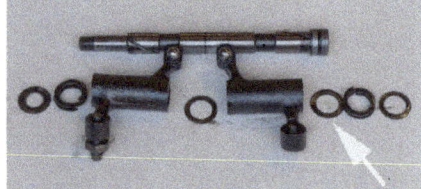

5.31 The rocker parts laid out – although I must have had a senior moment when doing it, as the washer arrowed should be with its partner in the centre, making the layout symmetrical.

5.25 A quick check on the top-end components removed so far revealed items like this: normal wear on the tappet adjuster.

ENGINE AND GEARBOX

5.32 The arms were put back on the shaft one at a time, and rocked to assess wear; they all came up fine.

5.37 When reassembling the rocker shaft, the new oil seal may get trapped on the way back in. This is the special tool designed to prevent that. It is inexpensive.

5.33 The remains of the old gasket took some scraping to remove, but, once off, the box was ready for cleaning.

5.38 Here it is, in place. Unfortunately, I did not find it particularly effective, as a small amount of seal was shaved off as the shaft went back in.

5.34 If any studs had needed replacing, the best way would be with two nuts locked together. The lower one could then be unwound to lift the stud out. Refitting can be achieved by reversing this process.

5.39 A new valve adjusting screw: this was a standard one, but hex versions for allen key adjustment are on the market, as are some with a mushroomed profile – supposed to lessen the load and thereby reduce wear.

5.35 Badly damaged studs may require the use of a dedicated removing tool. This type is fine for smaller studs.

5.36 Cylinder studs would need something a bit heftier, such as this half-inch drive remover.

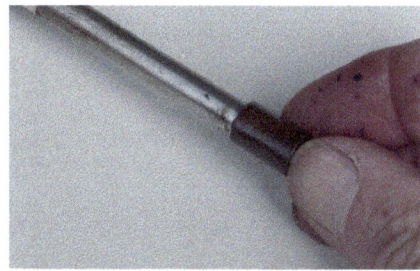

5.40 Whilst working on the valve train, I decided to check out the pushrods. These were rolled on a flat surface to check straightness, then the cups were pulled and twisted to make sure they had not come loose.

57

HOW TO RESTORE TRIUMPH BONNEVILLE T140

Cylinder head

Having found the damage to the adjuster pin as shown in the pictures, the first port of call was the corresponding valve end, which was clearly distressed. All the valves were coming out anyway, so, compressor in hand, I tried to move the damaged one. Absolutely no joy, so out with the soft-faced hammer, and several vigorous taps later, still no joy. The top keeper was clearly stuck to the collets. Another round of belting eventually saw it free, and the spring was compressed. With the springs off, I pulled the valve a little way out its seat, and wiggled to assess guide wear, of which there was plenty. The valve, however, could not be removed completely, as the end of the stem had become mushroomed, so some

5.41 This valve showed the usual wear expected on the stem when it has been in contact with the adjuster.

5.42 This was the one that was under the damaged screw seen earlier. The stem top was recessed and the edge rolled.

5.43 The valve heads showed a marked difference in colour, and oil contamination.

5.44 The valves were removed using a compressing tool to release the collets. Each valve has two springs, a metal seat and a top collar.

5.45 Before cleaning and starting the overhaul, I used a straight edge to check for warpage. All looked good; the 750 motors do not suffer from this as badly as the 650s, apparently.

ENGINE AND GEARBOX

5.46 All threads were checked, including the sparkplugs, so any rectification work could be done now.

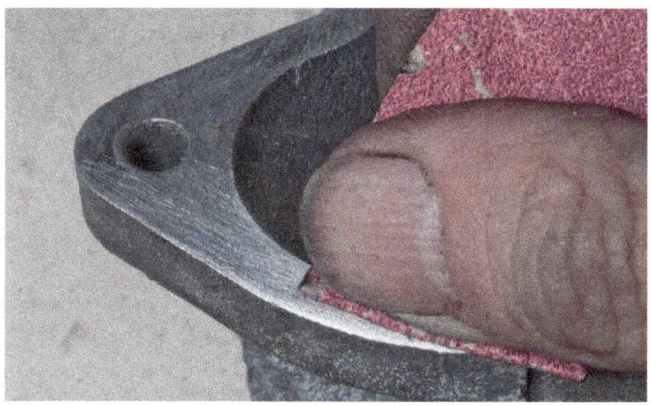

5.47 All gasket surfaces were thoroughly sanded and cleaned …

5.48 … and the remains of any old seals removed; in this case, from the pushrod tube recess …

5.49 … which was then cleaned out and sanded, as well.

5.50 The exhaust port was examined for signs of wear or damage from a loose fitting exhaust, but, once again, all was well.

5.51 Each valve was lifted just off its seat and wiggled to gauge wear.

5.52 Even the damaged valve did not have a really worn guide, but it was noticeably looser than the others.

5.53 Unfortunately, the mushroomed end prevented it passing through the guide. The only solution was to grind the end down until it would pass through.

5.54 With it out, the guide would have to come out, too. The head had to be clean before this was attempted, or the alloy surrounding the guide would have ended up scored and damaged.

5.55 The whole head was decarbonised, using a variety of plastic scrapers and very light abrasives, lubricated with carbon dissolving cleaners. It was a long, slow process.

work with a die grinder was needed before it eventually passed through. All the seats looked fine, as were the other guides, and the springs were measured and found to be within specification.

It is not unknown for Triumph heads to warp, although this was less of an issue on 750s, but better safe than sorry. So I used a straight edge for a quick visual check, followed by a brief lapping session on some 800s wet-and-dry, taped to a sheet of glass and lubricated with a little light oil. The mating surface was then checked for an even finish. Fortunately, mine came up okay. The area from the plug thread to the valve seats was examined for cracking, which is also not unknown. Happy with the condition of the casting, it was time to address the main issue.

If the guides need replacing, like mine, then it may be time to consider

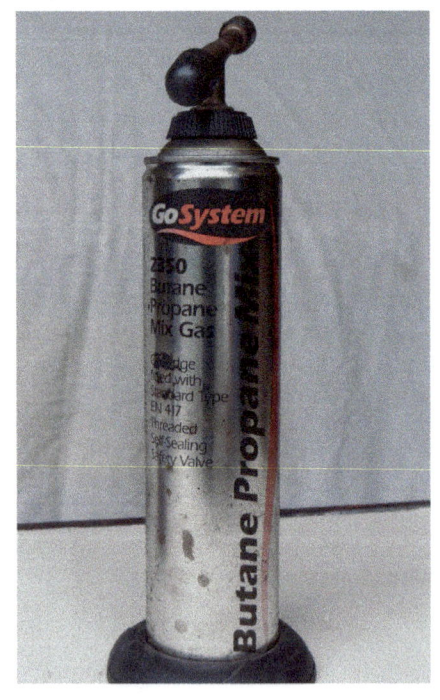

5.56 A gas torch like this, with a disposable cylinder, is ideal for warming castings.

ENGINE AND GEARBOX

5.57 The guide was then removed using another special tool, essentially just a stepped drift. The head was heated locally around the guide first, then it was hammered out. It came out cleanly, with no damage to it or the head recess.

5.58 The new guide and circlip: earlier models had a shouldered guide.

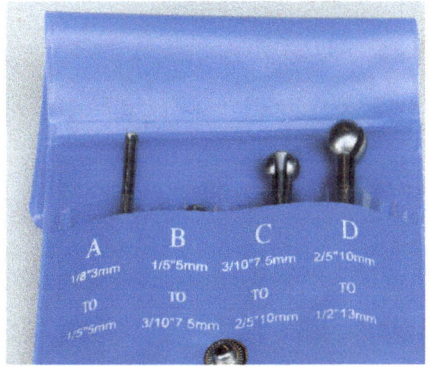

5.59 An alternative way of measuring guide wear, other than the wiggle test shown already, would be to use these expanding ball-headed tools. They should be inserted into the guide, the end twisted until the ball is in contact with the inside walls, locked, then removed and measured with a vernier gauge.

farming the work out to a machine shop, as manual methods carry some risks, including damage to the head. If you wish to do it yourself the old-fashioned way – and lots of people have done it over the years, including noted Triumph specialists – then the head has to be thoroughly cleaned, to make sure that you are not going to drag debris through as you drift the guides out, which would damage the relatively soft alloy of the casting. The area needed to be heated with a gas torch, or the whole head can be put in the oven (200°F for 45 minutes should be enough), and then the guides driven out using the dedicated punch tool, pictured. I was advised to drill through the guide for roughly three-quarters of its length, and use a parallel punch on the remains, as it was safer and less likely to cause damage, but I was concerned about keeping everything in line, so discounted that option, and went ahead just knocking them out, which took some effort.

Best practice would be to measure the hole in the head and the outside diameter of the new guide, to check the clearance, which was small: one to one-and-a-half thousands of an inch – something you are going to struggle to do accurately

5.61 A different type of expanding tool (discussed later) can be used to measure the hole in the head where the guide sat, and compared to the outside diameter of the new guide.

5.62 Before fitting the new guide, the hole must be thoroughly cleaned, in preparation. Once more, the alloy of the head was locally heated, and the guide, fresh from an overnight spell in the freezer, was hammered home until the circlip nestled in its recess.

5.60 Even budget electronic calipers should be sufficiently accurate for these measurements.

5.63 The stem and back of the head of the old valves, which were going to be reused, were cleaned on a bench-mounted wire brush.

61

HOW TO RESTORE TRIUMPH BONNEVILLE T140

5.64 Heavy deposits on the face were rubbed away on coarse production paper, before they, too, had a session on the wire brush.

5.65 The edges had minor marking, but nothing too heavy, and the stems were within tolerance when measured.

5.66 The faces cleaned up a treat, as well, so the three old valves went back in, with a new one to match the new guide.

5.67 Old-fashioned grinding paste works perfectly well, but, for a quicker job, there are diamond-based compounds, which work in a tenth of the time – for about ten times the price.

5.68 The grinding process was arm-achingly slow, but still worth the effort, in my opinion, despite having fallen out of favour in many quarters recently.

5.69 Most of these motors were not fitted with valve stem oil seals, although a few bikes at the end of production came with Norton-style guides and seals. One line of thought on my type of engine, is to fit Viton O-rings on the valve stem. Worth a try, especially as my local seal firm could not be bothered starting its computer for such a small amount, so I got them free.

5.70 With the spring compressed, the O-ring was slipped over the valve stem; then the collets were fitted. When the tension was released, the seal got trapped between the top collar and collets preventing any puddled oil dropping down the stem.

ENGINE AND GEARBOX

at home, without decent measuring equipment. If you don't have the gear, and there was good resistance coming out, and no debris on the guide, it would not be unreasonable to assume the new one would be a good fit. If the head had been worn, oversize guides are available, but professional advice would definitely be needed to accurately measure for the correct size. After guide replacement, the seats really needed to be re-cut to match, but you can always make a rough check, using fine grinding paste and seeing if the valve has a consistent contact area with its seat, although that isn't one hundred per cent foolproof. Seat cutting isn't really an area where the enthusiast restorer is able to do much; decent cutting tools are expensive, and hand cutting is a skill which would take time to master, and you could do a lot of damage whilst trying to acquire it. I used a small engineering shop which charged me only £10 a seat, so the expense need not be great.

New valves can be cheap, but make sure you buy them from a specialist, as poorly manufactured ones will not have their heads correctly aligned to the stem (as my machinist showed me on a previous rebuild of another air-cooled twin). Obviously, in that condition, it could not be matched to its seat and back it had to go.

My '79 T140E did not come with valve stem seals, nor did any bike leaving the factory until 1981, so these are pretty rare. Norman Hyde does aftermarket versions which come highly recommended. Another option is to use a ring seal on the stem. The theory is that most of the oil getting to the guide actually runs directly down the stem after it has collected in the top retainer, so the seal prevents it dropping down and into the guide.

PISTONS AND RINGS

The pistons were carbon-fouled and exhibited blow-by, which indicated that the rings, at least, were past their best. They were removed and the pistons cleaned on a bench grinder fitted with a soft wire brush, which was great for removing this sort of stuff without damage to the

5.71 The top and middle rings looked in good condition, sharp-edged with plenty of spring. The oil control rings, though, were quite the opposite.

5.72 A pair of ring pliers would be a handy addition to any tool box – more for fitting than removal, but still useful for the latter, as well.

5.73 The old rings came off easily enough.

5.74 The pistons cleaned up well to reveal their original part numbers and a standard bore.

underlying alloy. Once clean, the piston top revealed a set of standard pistons. A quick measure with a digital caliper confirmed their status, and that they were within factory tolerances, so they could go back in. The ring grooves were pretty clean, but varnished, so were lightly rubbed with fine wet-and-dry, after a wipe with thinners. Piston measurements and sizing are fully covered in the Factory Manual.

5.75 The ring grooves were cleaned. If they were really cruddy, a piece taken from an old ring could be used as a scraper. I finished off with wet-and-dry paper, lubricated with carb cleaner, as they must be spotless for the new rings.

5.76 The gudgeon pins were cleaned and, despite the camera showing every mark, they were, in fact, perfectly reusable.

5.77 I chose Harris rings, as they were recommended to me by several people. I was warned that some Far-Eastern rings are very poor and wear quickly, although pistons from the same source are perfectly alright. Such are the vagaries of the aftermarket.

HOW TO RESTORE TRIUMPH BONNEVILLE T140

BARREL

The barrel was cleaned and degreased to get a better idea of wear. The bores were free from scores or any other obvious damage, so they were measured internally to confirm their status. The correct position for this was indicated in the Factory Manual. Externally, the original paint had flaked off and rust formed, in some of the inner nooks and crannies, but had generally held up very well. If a complete refinish had been required, soda blasting would probably be the kindest method, and should provide a good surface for the application of heat-dispersing paint, although harder blast media, followed by powder coating, is also popular, despite some concerns about heat transfer. A brushing enamel designed for engines would be perfectly adequate for minor cosmetic touching up.

If there had been any major issues, other than a rebore, then new barrels are manufactured in the UK and available at a reasonable price.

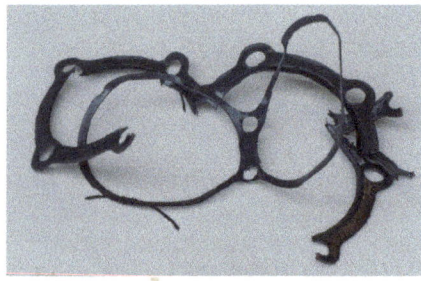

5.80 The head gasket wasn't any easier.

5.81 The bore was measured using an internal telescopic gauge, which locks in place. Two measurements were taken at right angles, to check ovality near the top and bottom of the barrels. A smaller version of this tool was used earlier, to check the hole size in the head, before fitting the new valve guide.

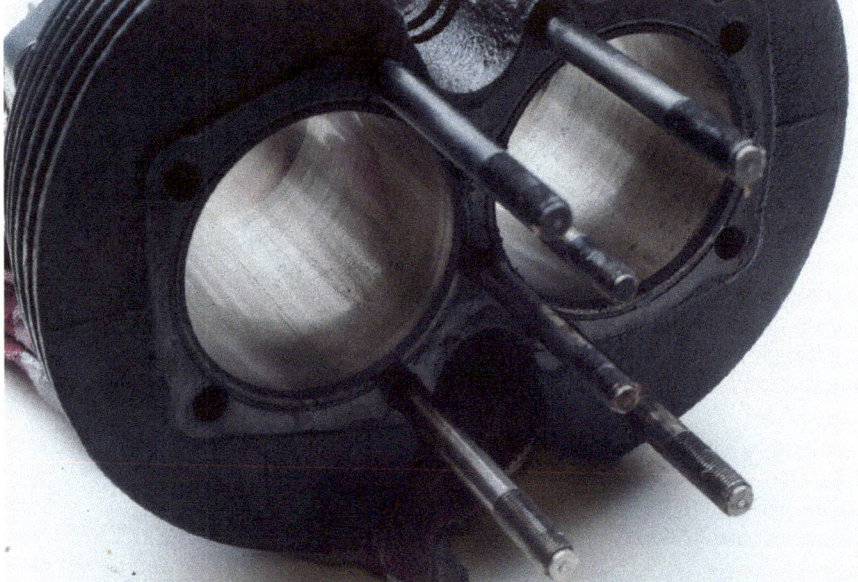

5.78 The bores had no major signs of wear or scoring, but looked as though they had been honed recently and not done many miles since.

5.79 The base gasket was really stubborn, and required the largest gasket scraper I owned.

5.82 The tappets and their blocks had to come out next. Under this metal ring there lurked a previous attempt at curing an oil leak: copious amounts of silicone sealer.

ENGINE AND GEARBOX

5.83 The tappets were a loose push-fit into the blocks. The bottom faces should be checked for wear, scoring, etc. These were absolutely fine.

5.89 The barrels appeared to have an enamelled finish and were generally in good condition. I decided to clean and then touch up the poorer parts with some heat-dispersant paint that was already on my workshop shelves.

5.84 The stems were also in good condition. These exhaust ones had the oil holes drilled in the side. Their orientation was carefully noted.

5.87 ... which was achieved using this special tool. It fitted snugly into the block, with the two prongs inside the tappet holes. It needed a few hearty taps from a hammer to drift the block out of the barrel ...

5.85 Each block was secured by a tapered bolt, which had to be removed.

5.88 ... a complete ring of old sealant came out with it.

5.86 Once the bolt was out, the block was ready to be driven out ...

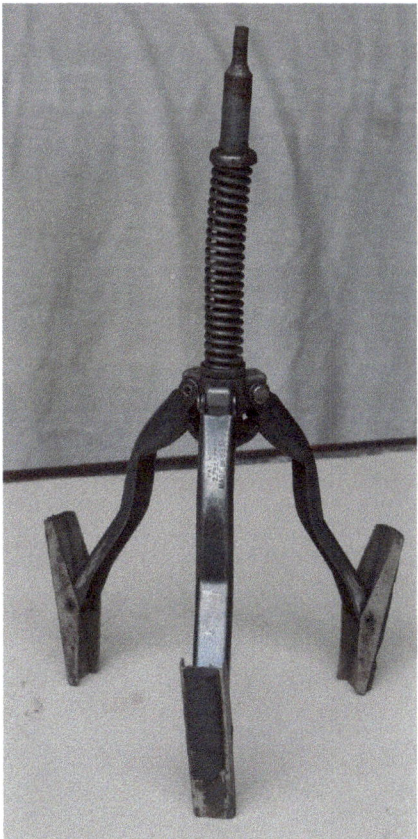

5.90 Before I applied it, though, the barrels were lightly honed with this tool. It only took a minute, but the abrasion would help the new rings bed in more effectively.

65

HOW TO RESTORE TRIUMPH BONNEVILLE T140

5.91 With everything back in black, the barrels were ready to be built back up.

5.96 The tappets were then oiled and reinserted.

5.97 The bottom of the pushrod tube had a red seal inside; it was removed with a slim pick.

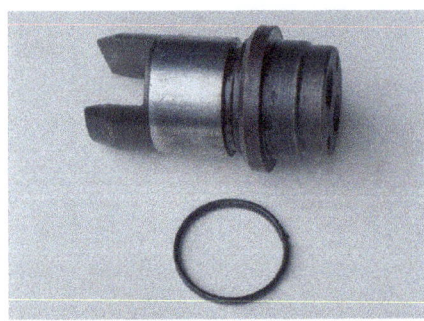

5.92 The blocks have an oil seal in a groove, and wear to this was likely to be the cause of the leak that the silicone had been used to staunch.

5.94 With a new seal fitted, the block was oiled and the hole in it, and the hole realigned with that of the barrel, before being drifted into place.

5.98 The top had an external black seal, which was in a poor state.

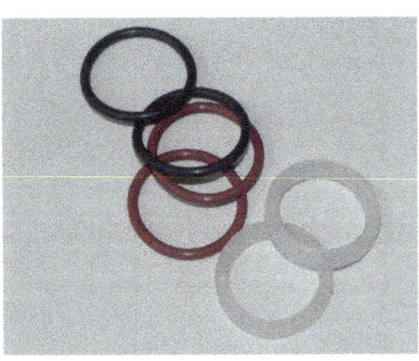

5.93 The full gasket set that I bought included all the necessary oil seals.

5.95 The translucent seal was then slipped over, followed by the steel band (called the 'wedding ring' in many circles).

5.99 Amazingly, the upper tube shown here started out looking as bad as the lower one, but responded well to degreasing, followed by some abrasive chrome polish. Replacements are readily available, if yours have proven to be too far gone.

ENGINE AND GEARBOX

ENGINE REMOVAL

With the top end off, and the mounting bolts out of the way, I lifted the engine onto the bench. Even in this partially stripped state, it was heavy, but I had planned my route to the bench carefully beforehand, so made it there unscathed. Dedicated stands are available to hold things steady for further stripping down, but wooden blocks, moved around as required, are perfectly acceptable, too, and, as a fully paid-up skinflint, that's what I used.

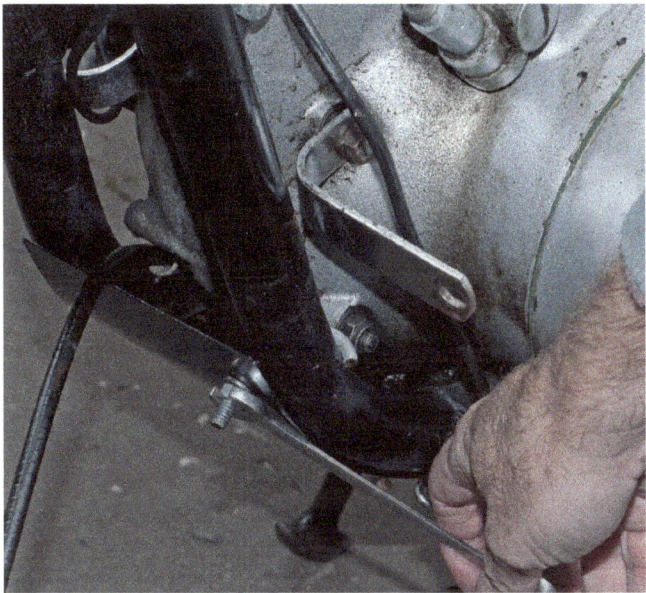

5.100 The non-standard bash plate had to be removed to gain access to the lower engine bolt.

5.102 The left rear engine plate was next: a straightforward nut-and-bolt job …

5.103 … as was the right-hand one. The bottom bolt under the engine was left to last. It had large and small spacers, like the front mount, with the same orientation.

5.101 The front mount was removed first, as it was easily accessible. Note the offset, thanks to the asymmetrical spacers.

5.104 The engine was extremely heavy and awkward to carry. It was only a short walk to the bench, but it seemed a lot further.

HOW TO RESTORE TRIUMPH BONNEVILLE T140

PRIMARY SIDE

If you have an impact wrench, having the motor in this state on the bench is not an issue, if you don't, then you will have to lock the crank with a bar through the con rods, as shown later in the picture sequence. I was surprised by the amount of oil left in there, despite draining, so a bit of mopping up was required. The strip down can be found in the photos, it differed from the manual slightly in order, but, apart from one slight error on my part, it worked.

5.108 With the cover off, the primary side was exposed, and it all looked to be in decent shape.

5.105 I had intended removing the primary side in its entirety before removing the engine, but changed my mind at the last minute. Now it was time to strip it.

5.109 The stator was the first piece to come off. It was held in place by three nuts on long studs.

5.110 The wiring sheath, where it exited the windings, was split and exposed, but the wire itself was undamaged, so it would be resealed, before it was put back in place.

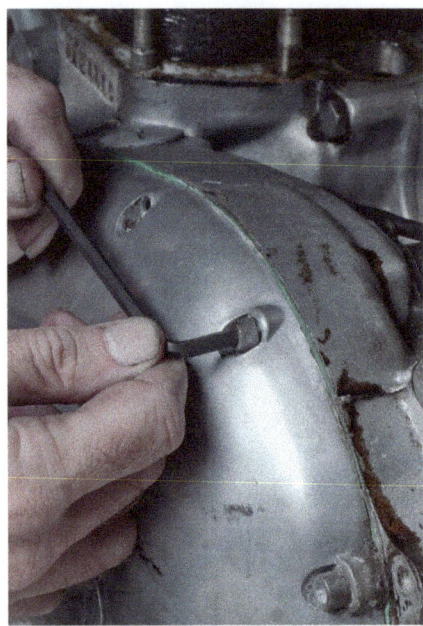

5.106 The outer case was held by allen-headed set screws and two nuts, all of which came off easily ...

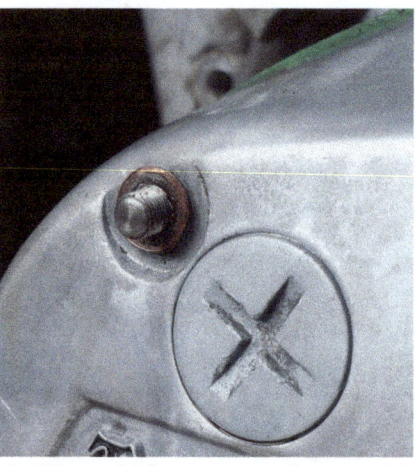

5.107 ... which was more than could be said for the two copper washers, which were a very tight fit. I was told to check the case, carefully, around these mounting studs, as over-tightening can crack them. These were fine. Had the case been damaged, remanufactured versions for '73-'78 models are available new.

5.111 The bullets on the three-way junction piece, on the outside of the case, were reluctant to come free, and had to be persuaded out.

ENGINE AND GEARBOX

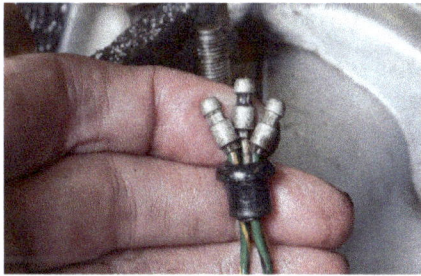

5.112 There was also a small rubber sleeve, where the wires entered the primary case, which was so hardened by age and exposure to hot oil, that it was no longer able to stretch over the bullets. The wiring was too hard to bend enough to try freeing them one at a time. I cut the sleeve, and would replace it later.

5.113 There was yet another rubber sleeve inside the cases, over the exit tube for the wiring; it, too, was rock hard.

5.114 But at least it could stay where it was for the moment, as the wiring was drawn back through the case.

5.115 The mounting nuts were set well back, so a deep socket was called for.

5.116 The stator was then lifted away. The rotor is held on by a large central nut, which may have a lock washer fitted. Mine had a simple shake-proof one.

5.117 With the nut undone using an impact gun (see later for alternative methods), the rotor simply pulled from the crank.

5.118 Underneath, there was a long square key, which was removed.

5.119 The large circular spacer behind could then be lifted off the crank.

69

HOW TO RESTORE TRIUMPH BONNEVILLE T140

5.120 The rotor and stator were immediately reunited, as this helped preserve magnetism. They were stored in a sealed plastic bag, otherwise they would have attracted metallic dust and debris.

5.121 The primary chain adjuster plug can be removed, and the tension taken off the chain, by undoing the adjuster in an anti-clockwise direction. A large, flat-bladed screwdriver was sufficient, although a special tool can be purchased.

5.122 The clutch had three nuts, and a central adjuster and locknut.

5.123 Each clutch nut had a peened edge. Each cup had one, too, that sat in a hole in the cover. A notched screwdriver could be fabricated or a tool purchased, but I managed to get them undone, using a large screwdriver in one side only …

5.124 … I remembered, later, that I actually had something in the toolbox which would have done the job. At least I could use it on reassembly.

5.125 Here, the nuts and their springs have been removed, as has the central adjuster. The cover is ready to come off.

5.126 This was what greeted me when it was removed. It all looked a bit mucky.

5.127 The first plain plate was removed to reveal the first friction one. Things appeared to be going well …

5.128 … and there it stopped: every other plate in the basket was stuck together. Not simply the type of sticking that needed freeing off each morning, but full-blown adhesion, that required a screwdriver and some force to separate. No wonder the clutch had felt heavy!

5.129 I kept everything together, in the order that it had come out, for the time being.

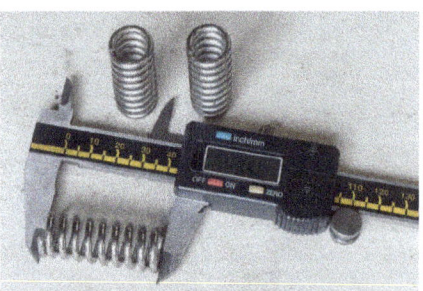

5.130 The old clutch springs were measured, in case I had to order new ones, but they were fine, and suspiciously clean, compared to the rest of the clutch components.

5.131 The clutch pushrod was pulled out next, followed by the centre nut, with the impact gun proving invaluable once more.

ENGINE AND GEARBOX

5.132 There was a washer under the nut, and it needed a magnetic pick-up tool to get it out, as access was limited.

5.136 With the puller having done its work, and the clutch centre out of the way, only the outer basket was left.

5.133 Hidden in the recess were some threads, and this was how the clutch basket was going to come out.

5.134 Another special tool, the clutch puller, and an unavoidable purchase: it was the only safe way to get the clutch off.

5.137 The crank sprocket needed yet another special tool. I pulled the clutch basket off at the same time, which allowed clutch rollers to scatter across the bench. I then read the manual and realised that, if I had followed the correct sequence, this could have been avoided.

5.135 Once screwed fully home, the centre bolt was tightened, and off popped the clutch centre without any problem.

5.138 The cross-over shaft (left-foot shift bikes only) had to be pulled out a little way, along with the clutch basket. It should come out easily, as it is located in a very simple fashion on the other side of the box.

71

HOW TO RESTORE TRIUMPH BONNEVILLE T140

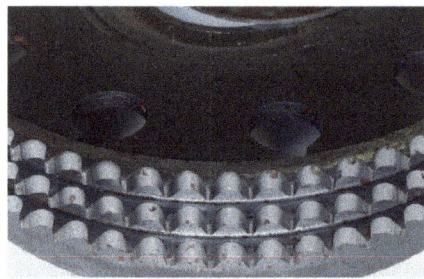
5.139 The teeth on the clutch basket were checked for wear, but were fine.

5.143 At the clutch end, the centre drive hub was drawn off its shaft; it was located by a Woodruff key.

5.147 Behind the cover lay the front sprocket, but its removal could wait for the moment.

5.140 The primary chain should be pretty long-lasting, but can be measured to check the degree of wear.

5.144 The gearchange through-shaft, which I had partially withdrawn whilst removing the clutch outer assembly, was then withdrawn.

5.148 In the top of the case was a plate, and the tube that the alternator wiring had passed through.

5.141 Under the drive sprocket, there should be a spacer.

5.145 The Woodruff key had to come out before the cover could be removed. It was very tight and took a fair amount of persuasion, so a replacement would be sourced for the rebuild.

5.149 The tube was actually a sleeve nut and was removed using a deep socket.

5.142 Under mine, there was also a washer, presumably added at time of manufacture, to help line up the drive chain so that it ran straight.

5.146 The plate was then unscrewed, and I took note of the orientation of the oil seal, which was not positioned as I would have expected it to be.

5.150 At the rear of the plate, screwed into the case, was the engine breather outlet pipe, which was removed by undoing the two screws.

ENGINE AND GEARBOX

5.151 The primary chain adjuster was the last to come out.

5.152 The adjusting nut was fully unscrewed and removed from the case.

5.153 The rod and blade could then be wiggled free from its locating stud.

5.154 The clutch centre had obviously been taken apart once before to get at the rubbers within, as the screws were seriously mangled. Fortunately, an impact driver got them out.

5.155 With the covers removed, the six rubber segments were exposed and looked like new. These usually wear quite quickly, so, once again, it did not look like the bike had covered many miles since work had been done on it.

5.156 The rubbers can be hard to squeeze back in, so I would have fabricated a tool similar to this one made for another bike, if I had to change them. It twisted the inner section, whilst the outer was lightly gripped in a vice. It worked a treat, before, and I am sure it would again.

HOW TO RESTORE TRIUMPH BONNEVILLE T140

TIMING SIDE

The timing side main cover was a little sticky, and needed gentle tapping with a plastic hammer to persuade it to come free. Once off, it was clear that the crank oil seal had picked up on the shaft – a good reason for not running an unknown motor, as pressure would have suffered as a result. The oil pump was unbolted, and it, too, took a little persuasion to pull off its studs.

Before removing the various gear wheels, I checked that it was all assembled correctly, and that the timing marks lined up. After several near misses, I checked the manual, only to discover that the engine might have to be revolved up to 94 times before everything lined up correctly, and I am pretty sure I was hitting that figure, before my paint-enhanced dots and dashes finally coincided.

The crank pinion nut, 1⅛in, was undone, and the Woodruff key tapped out. The special tool needed for the pinion nut had been made by the same firm that had produced the base nut spanner, and, once again, was not up to the job; at least not straight out of the packet. 15 minutes

5.157 Timing side removal started with removing the finned points cover.

5.158 My 'E' model was fitted with electronic ignition, although some still retained the points set-up from earlier models. The wiring connectors proved difficult to get a grip on.

5.159 It proved easier to remove the two pillar bolts and allow the plate to hang down.

5.160 The bolt in the centre of the reluctor (basically a spinning plate) was undone. The reluctor itself was on a taper and there was a tool to remove it. However, it was not tight; I inserted a small drift into the bolt hole, and a couple of very gentle taps saw it free.

5.161 The timing cover was then removed, and laid to one side for the moment …

5.162 … which revealed the timing gear set-up.

5.163 The oil pump was unbolted, and it was pulled reluctantly from its studs, along with the square cam plate which drives it.

5.164 I then rotated the engine, and lined up the timing marks, which I had highlighted in white paint. The crank pinion nut was removed with an impact gun, and the intermediate wheel, above, pulled off to increase access.

5.165 The crank pinion needs another special tool to get it off. Not cheap, but absolutely essential.

ENGINE AND GEARBOX

5.166 Unfortunately, it was unfit for purpose, as supplied, as the casting was far too thick to sit behind the pinion. Repeated bouts of grinding and checking eventually got it to fit.

5.167 The pinion came off easily, using the tool, to reveal a washer which had been misaligned on assembly, and trapped under the Woodruff key.

5.168 There was a thicker spacer, as well.

5.169 The oil pump gasket was removed; then the unit thoroughly cleaned.

5.170 The pistons showed minor marking (again, the camera made things look worse than with the naked eye).

5.171 With everything clean, the unit was lightly oiled; then the holes covered with my fingers, as I tried to pull the pistons back out – which proved impossible, and confirmed that the pump would draw correctly. It can be disassembled further to re-seat the balls inside, but I think I would have gone for a replacement, if it had failed either the visual or pull-test I carried out.

5.172 The drive block was checked for fit on the cam spindle …

5.173 … and the edges checked for wear or damage.

5.174 Back, then, to the timing cover, where the bottom oil seal was secured by a circlip, which was removed, and the seal pulled out.

5.175 The top seal just pushes out. Note the orientation, as it differs from the lower one.

on the grinder, broken up by repeated trips to check clearances, and it finally went on and did its job without drama.

GEARBOX STRIP

The Workshop Manual described gearbox disassembly in a remarkably short section, which made me slightly nervous. The gearbox cover was held by allen-headed set screws (3/16) and two nuts (½in), which were removed. Then the kickstart lever was turned, and the case gently tapped, and off it came, along with quite a lot of oil, despite the box having been drained earlier. The inner cover pulled away quite easily to reveal the gearbox. I fully intended to follow the instructions in the manual, as I am more used to horizontally split engines with the box in the lower crankcase, but it looked straightforward, so I leapt in and just started pulling gears out one by one, having first taken out the selector rod, with only two real issues. First, the circlip on the layshaft has very small eyes, and was tight, so I struggled getting it off, relying on piston ring pliers to spread it in the end and

HOW TO RESTORE TRIUMPH BONNEVILLE T140

5.176 The gearbox outer cover was held by a variety of fittings.

pulling it down the shaft. The second issue, which became obvious after I had done the first one, was that the selectors have to be lifted up to disengage the rollers from their slots in the cam plate, before the attached gear can be withdrawn.

The gear change cam plate slides out once the high gear assembly has been removed, which is discussed later, but the detent unit, 1⅞in, has to be unscrewed from the underside of the case, first. With everything out of the case, it was all cleaned in the parts washer, and inspected for damage. It was in excellent shape and fit for re-use, which was a relief. I then re-read the manual to double-check that I had not inadvertently done anything stupid during the strip (which I hadn't), so taking it apart was very straightforward.

Inner cover
Stripping the inner cover and changing bearings was straightforward, and is covered in the photo sequence

Outer cover
The main issue with the outer cover was the bent and damaged cotter pin holding on the kickstart lever. The other components came apart easily and as per the book.

5.177 The gearbox had been drained before the engine was taken out of the frame. The drain plug was the larger of the two nuts (arrowed).

5.179 The smaller plug blocked off this tube, which acted as the gearbox oil-level checker. The copper washers came in the gasket set.

5.178 I noticed that both cases were stamped, next to the drain plug, and the numbers matched, so at least the case halves had left the factory together.

5.180 With the outer cover off, the inner was exposed.

76

ENGINE AND GEARBOX

5.181 One of my fixings was a large cross-head screw, which, thankfully, was not too tight.

5.185 The rubber oil pipes and their metal junction box were cleaned and then removed.

5.182 The kickstart quadrant nut is secured with a tab washer, which was undone, and the trusty impact gun used to free the nut.

5.186 With the inner cover pulled free, there was a loose washer and a cog, which were removed ...

5.183 It then popped out under spring pressure.

5.184 The quadrant and spring lifted away together, like this.

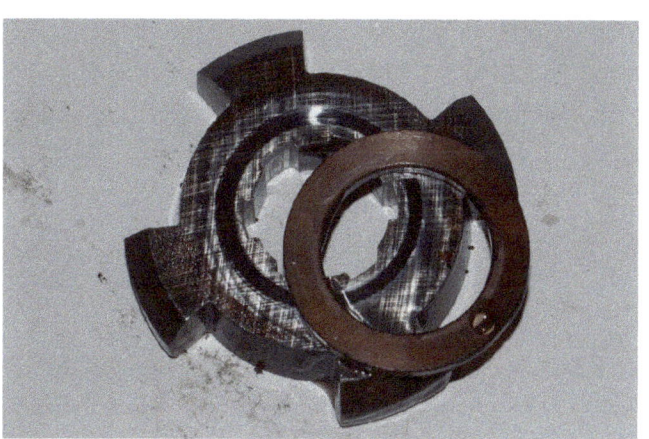

5.187 ... and looked like this.

5.188 Next to come off was this circlip. It proved to be really tough, so a decent set of pliers was required.

5.189 The tube (selector fork rod) that held the selector forks was then withdrawn.

5.190 The nearest selector was lifted up, which freed it from its groove, and it, along with the lowest gear, was taken out.

5.191 I then worked my way down through the box, alternating bottom to top, carefully laying out everything I removed, in order.

5.192 These three came out as one.

5.193 I kept going, lifting the selectors out of their track; it all came out very quickly until the shafts themselves were free and could be removed.

ENGINE AND GEARBOX

5.194 There were thrust washers under the shafts, which should be removed and measured/replaced.

5.195 The upper shaft runs in the high gear assembly, which would stay where it was, until I was ready to strip the cases themselves.

5.196 The gears run on these bushes. Each one should be slid back over its shaft and rocked to check for wear. These were absolutely fine.

5.197 These dogs at the edge of the gears take a bashing and their edges round off. This one had some signs of wear, but not enough to make me consider replacing it.

5.198 The edges of the gear teeth should be closely examined, as they, too, lose their edges. Once again, mine were in good condition.

5.199 The flat edges, and inside the teeth, may have pitting or flaking, if the case hardening has begun to wear.

5.200 The index plunger was unscrewed from the bottom of the gearbox. It sat next to the drain plug.

5.201 The sides of the plunger should be clean and not badly scored, and it should obviously be free to move up and down in its sleeve.

5.202 The business end that touches the gearchange cam plate can wear flat. Once again, mine was okay.

5.203 The tracks should have clean, sharp edges. When worn, they round off. This was just starting in a couple of very localised areas on this one, but nothing to worry about yet.

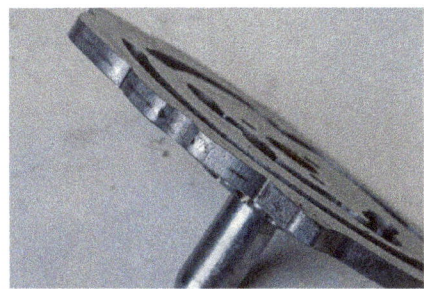

5.204 The index plunger tip ran on this edge: you can make out witness marks, which at this stage were really just discolouration rather than grooves.

5.205 The edges of the selectors can also round off, giving poor gear selection and retention.

5.206 The bottom shaft sat in these two needle roller bearings.

5.207 They were blanked at the rear of the case.

5.208 The only way to get them out, or, more importantly, back in, was with a suitably stepped drift, as they were delicate. On reinsertion, apply force only to the side of the bearing with numbers on.

5.209 The kickstart cotter pin was an issue, as it was bent and would not come out. I resorted to unscrewing the nut a little, then using a punch to try and square the shaft …

5.210 … it came out eventually, although obviously was not reusable.

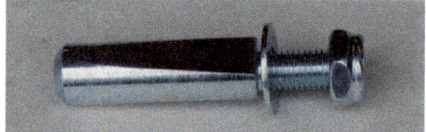
5.211 The benefit of choosing a popular British bike to restore is that parts back-up was excellent, so a replacement pin was no problem.

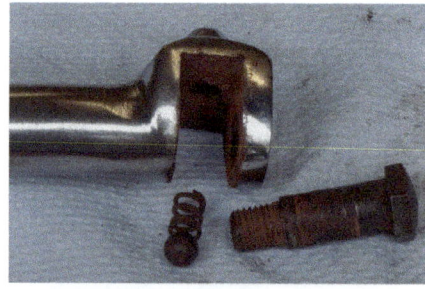
5.212 The swivelling top of the kickstart didn't swivel, as the ball was rusted in its hole, so it was picked out for cleaning and lubricating.

5.213 The rubber was completely fossilised, so was cut off, and the metal pin cleaned and painted.

ENGINE AND GEARBOX

5.214 Inside the gearbox outer cover lay the kickstart quadrant.

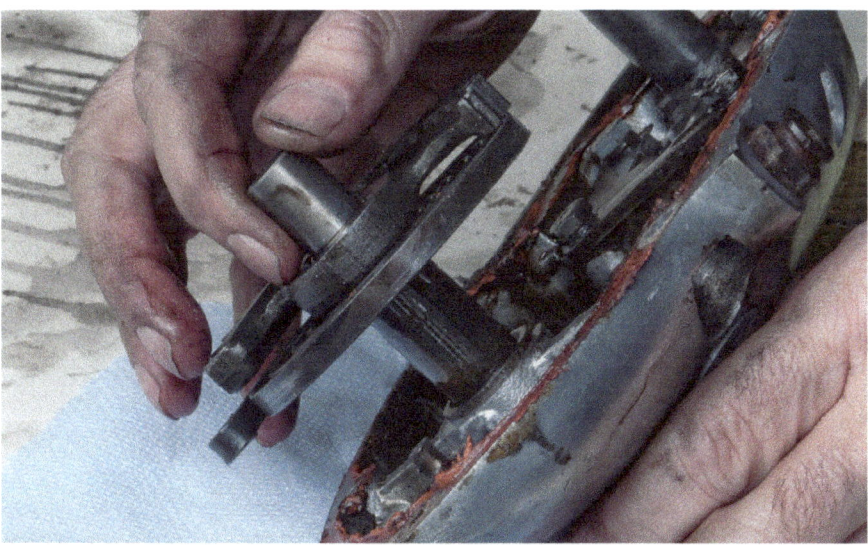

5.215 The spring was unhooked, and the shaft pulled out of the case …

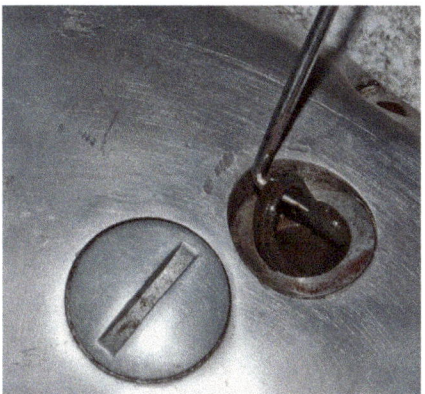

5.216 … in order to replace the oil seal, which is a common source of leaks.

5.217 When preparing the case for cleaning, I noticed that the alloy around the spring post was damaged. I couldn't work out what had caused it, but the pin was secure so I left it as it was.

5.218 Someone had been in here before, and obviously loved their gasket cement. It was all cleaned off.

5.219 Next up were the quadrant plungers under their guide plate.

5.220 Before stripping, I noted their orientation, as they are often assembled incorrectly.

5.221 Removing the plate would have been easy were it not for this bolt, the head of which is too close to the case to get a socket on. It wasn't much easier with a spanner, but it was eventually released.

HOW TO RESTORE TRIUMPH BONNEVILLE T140

5.222 Under the plate were two springs and the two plungers now ready for removal.

5.223 The plunger sides needed to be free from scoring or other damage.

5.224 The two springs can often break, and would also weaken with age and mileage.

5.225 The main quadrant itself was examined for wear …

5.226 … as was the bush it sat in.

5.227 The clutch lifting mechanism sat next to the quadrant assembly.

5.228 It has a simple ramp mechanism to convert lift from the clutch cable to push for the operating rod. As long as it was securely mounted, and moved freely through its arc, there wasn't a lot else to check. If it wasn't any good, it simply unscrewed for replacement.

ENGINE AND GEARBOX

BOTTOM END STRIP
Tachometer drive

The unit had a slotted cover, which simply unscrewed to access the internal workings. Underneath, there was the drive spindle, which can be withdrawn using a magnet to give access to a 7/16 nut which had a left-hand thread. Once undone, the whole assembly came away ready for cleaning and inspection of the drive teeth.

5.229 The tacho drive had a slotted cover …

5.230 … under which lay the drive spindle. A magnet was useful for drawing it out.

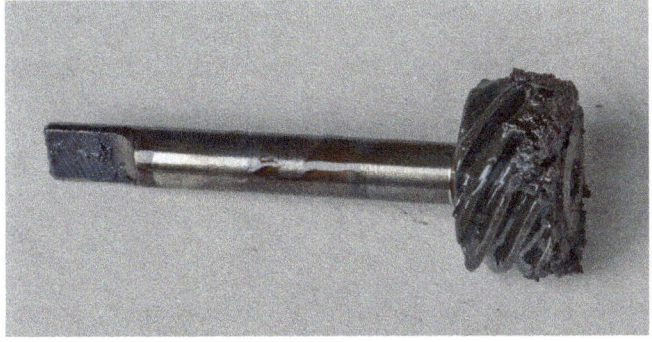

5.231 It was a bit cruddy, but the drive spiral was fine. If you want to remove the housing, there is a nut in the spindle recess. It has a left-hand thread.

Oil pressure unit

The oil pressure cap nut, 15/16in, simply unscrewed, and the main body followed, by undoing the inner nut 1in. The sleeve inside was checked for scoring or other damage.

5.232 The oil pressure relief unit is on the crank cases (arrowed).

5.233 It was stripped and cleaned. The main issues would normally be a weak spring and damage to the piston.

Splitting the cases

With the cases stripped of all ancillary bits, the remaining fittings were undone, 9/16in nuts and 1/2in through bolts and nuts. The back of the primary side case was gently tapped with a plastic hammer, but nothing budged. A double check that there wasn't a hidden bolt I had forgotten revealed nothing, so the crank was tapped, again with the plastic hammer. This produced a small amount of movement, then it all stopped again. There is a dowel at the front engine mount which can corrode, but mine was bright, shiny and loose. I was unable to get any further movement tapping on the primary side, so a tubular drift was found which would fit over the crank protrusion on the timing side, and this was used to apply some direct (but, once again, gentle) effort which saw another small amount of movement; enough, this time, to insert a slim wooden wedge into the front of the cases. Alternating, then, between tapping the rear and pushing the wedge further into the case gap saw everything apart. I was expecting the primary side bearing to have separated,

5.234 The case bolts were removed and the splitting process started. It was slow going, and, at one point, I went very retro, using wooden wedges to persuade movement without risking damage to the alloy.

5.235 When split, they left the crank ready for removal.

ENGINE AND GEARBOX

and the inner to be on the crank, but it had stayed in one piece in the left-hand case. If it had remained on the crank, it could be removed using the same process as the steering head bearing mentioned in a previous chapter.

CRANKSHAFT

The conrod nuts (½in) were removed to allow the bottom caps to come off, which needed a few taps with the plastic hammer to get them moving. The big end shells were in perfect condition, as were the crank journals, which were measured with a micrometer just to double check. Three undersizes are available if required; but ensure that your chosen engineering shop for the re-grind is used to working on motorcycles, as the journals are shouldered (not parallel like many car applications) and, if incorrectly done, can crack the crank.

The picture sequence of the crank sludge trap plug does not include the difficulties of refitting, which are noted here: a trial fit of the new plug saw it tighten up almost immediately after insertion – so much so, that it was obvious it would never make it all the way down. The crank threads were carefully inspected, but seemed to be in excellent shape. Nonetheless, I cut a couple of slots in the old plug, and it was worked in-and-out with lots of lubricant, to make doubly sure that all was clear. Any problems remaining would obviously be with the new plug, which was given a dab of Loctite and started off on its threads. This time it went in further, so, now confident that it would make it all the way, I kept winding it down, but it was tough going for the last couple of turns. It was finally peened over, using a centre punch to stake it in place, making sure that it was actually doing its job this time around.

CAMSHAFTS

The camshafts simply pulled free from their bushes on the primary side, which were in perfect condition. I was told that they don't wear quickly, but in the event that they have, removal of the blind pair would be tricky, as, without a dedicated puller set, a coarse threaded tap should be wound

5.236 The big end cap nuts were removed. They cannot be reused, but were saved for the time being, to keep the rods together until the rebuild.

5.237 The shells were in really good condition without any scoring or other marks. They were removed by gently pushing on one edge; they then slid round and out.

5.238 The rods should be marked from the factory – mine were. If not, I would have made my own marks, so that the caps remained with their rod and mounted the right way round.

5.239 The journals were fine, too. I noted the shoulders, to the flywheel sides.

5.240 Each of the three flat faces were closely examined for wear or other damage, as were the threads and the Woodruff key slot. The smallest section on the left was critical, as it sat in the timing side oil seal, and any problems would mean that the oil pressure was affected. It could have been reground and an oversize seal used.

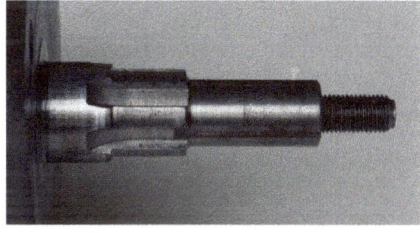

5.241 The same checks were made on the other side of the crank.

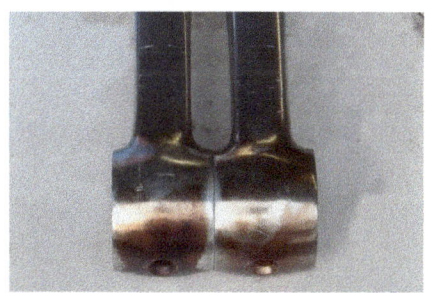

5.242 The rods were laid side-by-side to check for twisting. They were then slowly turned against each other for signs of gapping.

5.243 Each gudgeon pin was put in its respective rod, to check for wear by trying to rock them.

HOW TO RESTORE TRIUMPH BONNEVILLE T140

5.244 The infamous sludge trap: the peen mark on mine looked as though it was contributing very little to the security of the plug, so I did not bother drilling it out.

5.245 I was convinced that my impact gun would make short work of spinning the plug out. I was wrong: it budged a tiny amount, then snapped the end off my bit, which had seen years of service taking out vehicle kingpin caps, so it was no weakling.

5.246 Further attempts with a variety of bits and screwdrivers only managed to chew up the slot, so time for extreme measures. The plug was cleaned up with a small grinder.

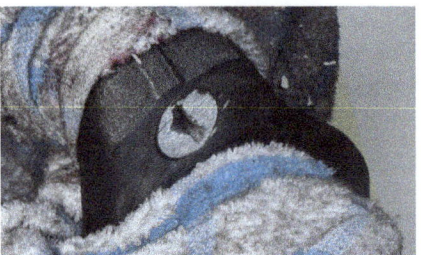

5.247 Then the rest of the crank was carefully covered up, in preparation for what was about to hit it.

5.248 A washer was welded to the plug to give sufficient width to then weld on a decent-sized nut. It still needed a ½in drive strong arm and lots of heaving to get the thing out …

5.249 … but it was worth the effort, as there was a reasonable amount of sludge in there.

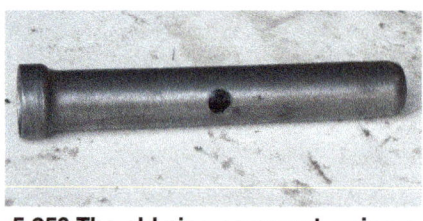

5.250 The old pipe came out, using a hook, and everything was thoroughly cleaned. I had bought a new tube anyway, as I expected the old one to be damaged in the removal process.

5.251 The new plug came with a far more sensible hex head, which offered much better grip than a screwdriver slot.

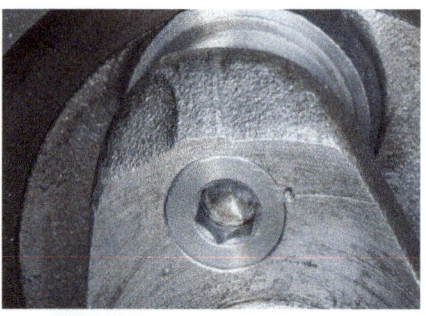

5.252 The crank threads were cleaned, and a little locking compound added to the plug. I used a larger punch in the original peened hole, to increase its size, so that it actually did something.

5.253 The cam lobes and bearing surfaces were all fine, with no sign of pitting or damage.

5.254 The bushes in the cases were fine, as well.

in and then trapped in a vice. The cases would then be tapped away from it. Given the cost of acquiring the necessary tools, it may be simpler just to get an engineering firm to remove them for you; it wouldn't cost much, and they can fit and ream (if required) the new ones at the same time.

ENGINE AND GEARBOX

GEARBOX HIGH GEAR
Although part of the box, I thought this would be easier to deal with once the cases were split.

CASE CLEANING
The main crankcases were cast alloy, and naturally had a rough finish, which traps dirt and easily stains from oil leaks. The best solution, with the least risk, would be to have them soda blasted. If, like me, home-based solutions are the preferred option, then a good soak in degreaser, and a blast from the pressure washer, gets rid of most marks. Very reluctant stains could be treated with a brief exposure to oven cleaner, but with care, as some formulations may attack the alloy. Professional compounds for dealing with alloy wheels work well, too. If all of that still leaves marks behind, then painting is the final option. I have used steel wheel silver aerosol with success in the past. I was also told that Simoniz engine paint gives a realistic finish.

REASSEMBLY
With all the main sub-assemblies stripped, checked, and with the problems addressed, it was time to bolt them all back together. Once again, it raised some issues where opinions differ. All the fittings that were torque sensitive had their threads left unlubricated, but were assembled with medium strength locking compound, although that in itself can alter the figures from dry. Their shanks were lightly greased to discourage corrosion. Some prefer to lightly oil the entire bolt before fitting.

Bottom end rebuild
The rods were placed back in the journals they came from, and bolted up using new nuts. Many recommend replacing the bolts as well, and, had my motor shown signs of high mileage-related wear, I would have done so, but it clearly didn't, and I was finally swayed by an article from many years ago, written by Harry Woolridge from the Triumph Service Department, who suggested the bolts could be re-used two or three times on a road-going engine. The new nuts were torqued down, and the rods checked for free movement, first by complete rotation, then by allowing

5.255 The reassembly of the bottom end was begun by pumping oil through the crank to make sure the oilways were clear. They had been blasted through with compressed air after the cleaning process, but better safe than sorry.

5.256 The drive side roller was seated with a hollow driver. The drift method shown in Chapter 2 could also be used, if done with care.

5.257 The timing side bearing was installed with a bearing tool into a pre-warmed case.

5.258 The bearings benefited from a smear of bearing fit compound before insertion.

each to drop to either side under their own weight, which they did smoothly, so all good there.

With the crank ready to go, I reinstalled the high gear bearing, followed by the gear cam plate, which has to go in now, then by the high gear assembly itself, and tightened up the securing nut. The only issue was the reluctance of the lock washer to clear the new O-ring oil seal, but it all went together after a couple of tries.

The drive side inner bearing was then drifted down on the crank, making sure the lip was up against the web. The crank was inserted into the timing side case, making sure that the rods were well protected and out of harm's way; then a couple of

5.259 The roller outer race was installed in the same way, on the other side.

5.260 The case drain plug had been cleaned and was reinserted, along with a new seal.

5.261 I usually use Rhodorseal, and have done for many years, but my supplier was out of stock when I needed it. This was suggested instead, and it proved ideal; and the built-in dispenser was useful. Many owners I spoke to recommended Wellseal.

5.262 A bead of sealant was applied to the case halves, and then they were reassembled, which went much more easily than taking them apart.

5.263 The cover plate got new seals and gasket, and the screws a dab of medium-strength locking compound.

ENGINE AND GEARBOX

5.264 Reassembly would require setting accurate, and often high, torque, so my impact gun would have to go back in the toolbox. The crank was locked by passing a rubber-sheathed extension bar through the little end eyes, which then rested on two blocks of wood to protect the cases.

5.265 The high gear assembly, which had been ignored during the gearbox strip, had to be addressed with the cases split, as it was more accessible like that. These pictures, and the procedure shown, must be completed before the cases are put back together, but it was easier to show the bottom end section all together, before I moved on to this. So … the gearbox sprocket was trapped in a vice, and the lock washer tapped back out of the way.

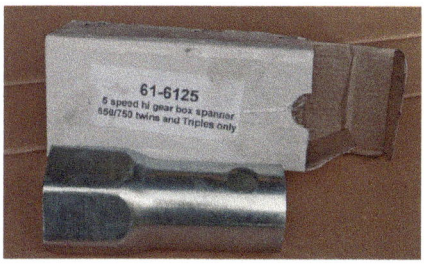

5.266 The nut was an unusual size, and, despite raiding a colleague's extensive selection of sockets in addition to mine, neither of us had anything that fitted, so I had to buy the special box spanner designed for the job.

5.267 Under the nut was a large rubber O-ring. When reassembling, this can be awkward to seat properly.

5.268 The sprocket was removed, and the central shaft tapped through, which was without drama. Underneath, the bearing was now clearly visible, along with a large oil seal around the outer edge.

5.269 The seal was levered out, to reveal the circlip securing the bearing. Removal and replacement was the same as previously: warm the case, and tap out, chill the bearing and reseat into a pre-heated case.

5.270 The high gear shaft was checked for chipping to the teeth, or other damage.

HOW TO RESTORE TRIUMPH BONNEVILLE T140

5.271 The lock washer and seals were an automatic replacement.

5.272 The needle bearings were fine in mine, but if they had required replacement, the stepped drift would have been needed, as with the gearbox needle rollers. The high gear shaft was inserted through the new bearing, and reassembled in reverse order without any issues, not forgetting to put the gearbox cam plate back in place, first.

decent taps with a soft hammer saw it seated. The joint face was coated with sealant, and the right-hand case carefully placed over the protruding stud and dropped on to the crank. It was then tapped down, using lots of little blows, again with the soft hammer, all around the case. It went on far more easily than it had come apart, and the securing nuts and bolts were tightened. The crank was rotated fully, to make sure everything moved smoothly, which it did.

Gearbox

Having taken the box apart in a non-factory manner, I decided to rebuild it with the same scant regard for the manual's advice. The cam plate was put into the neutral position and the plunger assembly screwed back into place. I checked that the plate moved freely; then selected first. The main shaft was then inserted with third and fourth gears on it. The selector fork was slid in from underneath, then slotted into its groove in the gear, and twisted anti-clockwise and inserted into its track, although I had to move the cam plate to get it in. I used thick grease to keep it there. The layshaft went in next, with its top two gears, then I worked backwards out of the box, fitting gears alternatively to top and bottom shafts, as I went. The selector forks were slipped on as I went; the two remaining ones were notched, so it was impossible to get them mixed up. Throughout this process, I had a print-out of the assembled shafts from the Workshop Manual next to me. The selector fork rod needed some assistance to get it in place, as the rearmost selector was drooping slightly, so a screwdriver was used to support it, whilst the rod went in. The circlip on the layshaft, which had been such a pain on the way out, was much easier to get back on, as, once spread, it could be slid back along the shaft into its locating groove. The driving dog went on last. Everything was given a squirt of oil from a can and the box was done – a lot more straightforward than I had expected. I ran through all the gears to make sure that it all worked, then selected neutral.

The gearbox inner cover was next up, which involved indexing the operating quadrant – I had been warned this could be troublesome. My bike, though, was fitted with a neutral switch, operated by a corresponding pip on the quadrant, and, as I had set the box in neutral, too, I expected it to all line up. A new gasket was fitted, then the cover was gently tapped down. The layshaft needed lifting slightly, using the end of a small screwdriver to line it up, and, once done, the cover slid on, until it was only held off by the interference of the quadrant and cam plate teeth – a very slight wiggle saw it home, with the quadrant sitting bang on in the neutral position relative to the switch button. Earlier models did not have this switch, in which case, the best option, I think, would be to put the box in first gear, and, with the operating quadrant allowed to dangle under its own weight, a slight upward movement, on fitting, should see it in the right position. There is a special tool to line everything up, but it can be done without it.

5.273 The gearbox rebuild began with a clean case. New bearings have been fitted and greased, ready to go.

5.274 The cam plate was put back in place before the high gear was fitted, as mentioned before, but now it was set in neutral, and the plunger reinserted to hold it.

5.275 The top shaft went in next.

ENGINE AND GEARBOX

5.276 The thrust washer for the lower shaft was stuck in place with grease …

5.277 … followed by the top selector and bottom shaft.

5.278 I then worked my way back out, inserting cogs and selectors, one at a time, alternating between shafts.

5.279 With the first two selectors in, I inserted the fork rod; which needed a little help from a screwdriver to line things up properly. The final cogs went back on, and the circlip followed to hold everything in place. Reassembly had been a straightforward reversal of my removal sequence.

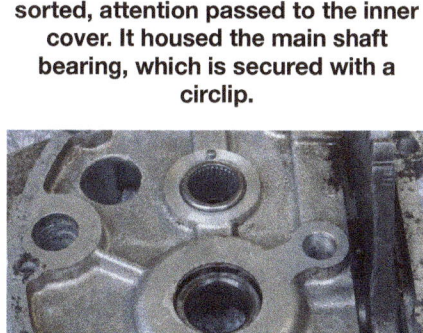

5.280 With the gearbox internals sorted, attention passed to the inner cover. It housed the main shaft bearing, which is secured with a circlip.

5.281 With the circlip removed, the bearing comes out from the rear.

5.282 With the case heated locally with a gas torch, it was tapped out using a socket as a drift.

5.283 The layshaft bearing requires a stepped drift, as used before, to remove and replace.

5.284 The new bearing and its washer were greased, before the case was offered up.

5.285 Getting the inner case back on was not straightforward, as the layshaft drooped and needed support, and the gearchange quadrant required jiggling very slightly, before it meshed and everything slid into place.

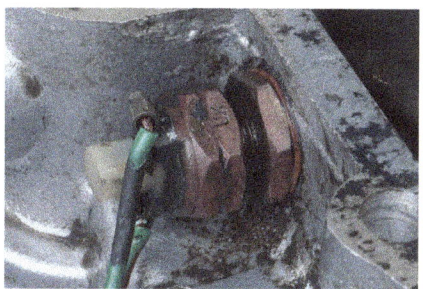

5.286 My box had a neutral switch, which was bench-tested, cleaned and reinserted.

HOW TO RESTORE TRIUMPH BONNEVILLE T140

5.287 The teeth of the kickstart quadrant were closely examined, as they chip easily.

Primary side

The only issue on reassembly was to make sure that the clutch inner and outer went on together, which I had foolishly missed during the strip down.

5.288 The clutch rollers were replaced, as they were inexpensive.

5.289 They were stuck to the clutch centre with grease, ready for the outer to be lowered over them.

5.290 The large thrust washer on the back of the basket was changed, too.

5.291 All the Woodruff keys were checked throughout. The clutch one had a ridged edge, so was changed.

5.292 The rotor seemed fine physically, so I carried out a rudimentary test to check its magnetism, by lifting it, using each segment in turn. It passed, so could go back on.

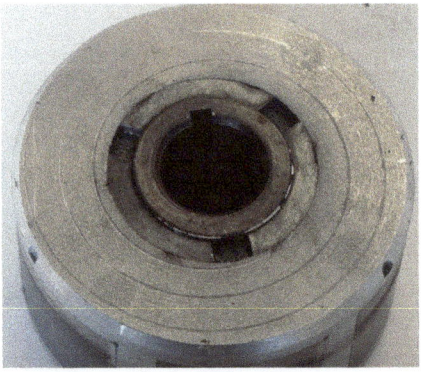

5.293 I was warned that the centre of the rotor sometimes works loose, but this one showed no signs of movement. I later found out that this affliction mainly affects earlier models.

5.294 A lesson learned from the earlier strip down: this time, I put the clutch assembly on complete, at the same time as the chain and rotor.

5.295 Another special tool: this time, a metal plate which bridges both inner and outer clutch parts, which allowed the centre nut to be torqued up. Very useful, and cheap.

5.296 The plain plates were checked to make sure that they were flat, then cleaned with production paper. The scratched surface that remained after cleaning helped to retain oil in the short term, but would soon smooth off in use.

ENGINE AND GEARBOX

5.297 The manual was unclear about the friction plates. For wet clutch bikes, they would be immersed in appropriate oil, like this, for a short time before assembly, but the Triumph case was not fully wet. I did not want to assemble them totally dry, so they were given a light smear before going in, to stop them sticking until the bike went into use.

5.298 The clutch pressure plate was checked for flatness, and the cups inserted.

5.299 The nuts were like new, so were reused. If they had shown any signs of sticking or corrosion when taken out, they would have been replaced.

5.300 The end of the clutch pushrod often burrs over.

5.301 Its length was checked, too.

5.302 The final job before refitting the outer case was to tension the primary chain. This time, an impact driver bit and a spanner were used, which was easier than a long screwdriver; and essential if the engine was back in the frame.

Timing side

As all good manuals say, reassembly was a reversal of disassembly, and that was pretty much true. I made certain that all the timing marks were properly aligned.

5.303 The timing side went back together in a straightforward manner. The only issue was to make sure all the timing marks were in alignment. The oil pump was carefully torqued, too, as it can distort otherwise.

5.304 The timing cover bottom seal was replaced. I was warned that this was not an area to try and save money, as cheaper seals are supposedly not able to handle the oil pressure, and sometimes flip, causing a potentially catastrophic drop in oil pressure.

5.305 A new points seal was pushed in. It takes a moment to get it square, as it is easy to push it too far and out the other side.

5.306 There is yet another special tool to get the shaft through the seal without damage. I didn't have it to hand, but, by slowly pushing the cover on and watching the seal, the area of contact was clearly visible and quite small, so I used a tiny electrical screwdriver to push the seal down, as the cover went fully home.

HOW TO RESTORE TRIUMPH BONNEVILLE T140

Barrels and pistons

My barrels were fine for re-use, so were honed and oiled. The new rings were lightly oiled on their contact surface only. Some prefer to really soak them with oil; others argue for a dry installation – there are pros and cons for all methods.

5.307 The pistons were reassembled with new rings, but I decided that I wasn't going to struggle with the original circlips, so bought some like these: originally designed for BSA singles (or so I was told). They were so much easier to fit.

5.308 The pistons were held parallel on two tyre levers. My ring clamps did not fit, so I used a combination of cable ties and my fingers to compress the rings into the barrels. A little fiddly, but ultimately successful. Next time, I will buy the right clamps.

5.309 For the barrel nuts, I found this old cranked metric spanner lurking in the garage. It looked like it might work.

5.310 It fitted perfectly, with just the right amount of crank to allow access to all the 12-point nuts.

5.311 The pushrod tubes were put in place …

Cylinder head

My T140E had a composite head gasket from new; previous models had a copper one, which can be re-used if annealed, but the mounting holes should be checked for signs of elongation first, and the gasket replaced, if there are any. Both types have their adherents, but I was quite happy to stick with the factory specification in my rebuild. For fitting the head, Service Bulletin 9/79 states that the metal bridge surface of the gasket fits towards the cylinder head. The torque setting for the 5/16 studs is 18ft-lbs and all 3/8 studs are 22ft-lbs, in the sequence laid out in the manual, which is what I did.

5.312 … followed by the head gasket, which was composite on the E models and later. The earlier copper version can be annealed and re-used, but I think I would be tempted to renew it.

ENGINE AND GEARBOX

5.313 The head was dropped on (carefully), and the head bolts inserted. The threads were left dry, but the shanks were oiled.

5.316 The rocker boxes were fitted. Aligning the pushrods to get them to sit properly was irritatingly awkward and required some care, as it was all too easy to think they were in place when they hadn't seated properly. I turned the engine over whilst holding each down in turn, to make sure all was well before going on with the rebuild.

5.317 New copper washers were in the gasket set for the rocker oil feed pipe.

5.314 The head bolts were then torqued down, in the sequence laid out in the manual.

5.318 The dome nuts were cleaned and the feed pipe fitted.

5.315 The rocker box gaskets were fitted once the pushrods had been dropped into place. Getting the cups through the holes took some care, to prevent splitting the gasket.

5.319 The head steady, which had been such a pain when the engine was coming out, had been thoroughly de-rusted, then painted. It would be well greased when bolted back on.

Chapter 6
Fuel and exhaust

FUEL TANK

The tank was held in place by a single bolt, hidden under a cap, and a bracing strap on the underside. With everything undone, mine was still reluctant to move, as it was a very snug fit on the isolating rubbers wrapped around the top tube of the frame. Internally, it was showing signs of rust and there was the staining at the rear, which I had initially taken for a leaking seam. If the tank had been unserviceable, new ones are on the market, by Harris, which are to original specification, and replicas from India. The latter have received a mixed response, which appears to be down to variable build quality, as they are reputed to be handmade, so they remain a bit of a gamble.

6.2 The other mounting components included a check strap at the front, a rubber buffer for the main bolt, and a blanking plug and cap for the tank tube hole.

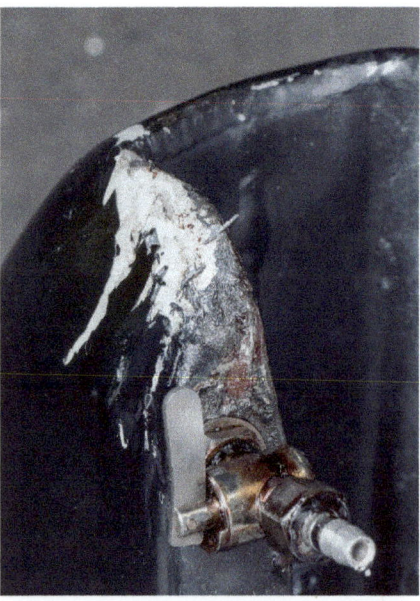

6.4 Missing paint at the rear indicated a leak, but from where precisely was it leaking?

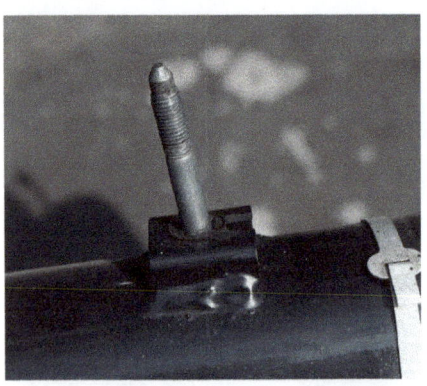

6.1 The petrol tank was secured by a central bolt which passed through a tube in the middle, a method that may seem unusual to those familiar with Japanese bikes of the era.

6.3 The frame had two of these dense foam buffers to isolate the tank from vibration. Everything was a snug fit, and my tank needed a decent pull to lift it free.

FUEL AND EXHAUST

FUEL TAPS

These were of simple construction. Sadly, there were no rebuild kits available when I looked, but a strip-and-check showed them to be gummed up, but serviceable. The circlip holding the internals was a pain to remove, as the lever seemed to get in the way, regardless of the angle of attack. If you cannot be bothered to struggle with them, new units are plentiful, but avoid the very cheap ones.

6.5 The petrol taps had clearly been leaking for some time, as their bodies were covered in brown sticky residue from evaporated fuel. The internal filters were missing, on one side, and damaged, on the other.

6.6 There was a Dowty washer between the tank and each tap body. It looked as though the old one had been reused after the tank had been repainted; this was the likely source of the leak. The situation was not helped by the absence of the spreading washer which should go between the nut and the Dowty.

6.7 The tap lever was held in place by a circlip which was very difficult to get pliers onto. It was eventually freed, after much cursing.

6.8 The tap then pulled out to reveal the rubber seal. I could not find a source for these seals, so, if it had been damaged, then new taps would have been needed.

6.9 Once cleaned, though, it proved to be in good condition …

6.10 … unlike the filter, which was distorted and had damaged mesh.

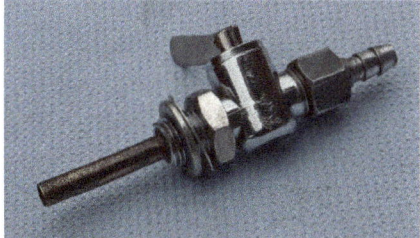

6.11 The taps cleaned well and, once fitted with the correct new washers, were good for a few more miles.

6.12 The inside of the tank was clearly corroded, and was even worse in the lower reaches where old fuel had sat for several years. Given the evidence of a possible leak, as well as the rust, I decided to seal it. – Better safe than sorry.

TANK SEALANT

Given the obvious corrosion and potential leak, my tank needed de-rusting and sealing to protect it from the ethanol in modern fuel, and there were several formulations to choose from to do just that job. My choice was Tank Cure, and the pictures show the use of that product, although most of them follow similar procedures.

6.13 There are plenty of potential sealing agents on the market: this was the one I chose, on the basis of a conversation I had with its producers, at an autojumble.

HOW TO RESTORE TRIUMPH BONNEVILLE T140

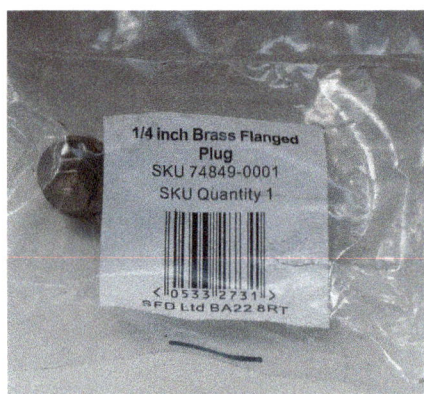

6.14 The tank orifices obviously had to be blocked during the process. Fortunately, being British, the tank threads were common to domestic plumbing fittings, so a pair of brass plugs were bought to seal the tap holes.

6.15 I also used PTFE tape around the threads, to make absolutely sure that there were no unwanted leaks. If I had been retaining the original paint, the whole tank would also have been carefully covered, as the sealant sticks to any surface with which it comes into contact.

6.16 The first part of the process was washing out with the supplied detergent. It was a slow job, with lots of sloshing around, and it took many rinses to get rid of the foam that was generated.

6.17 After washing, the tank had to be dried, so I used a hot air gun, although a hair drier could also have been used. I had to take care, as the tank became very hot when I lingered with the gun.

6.18 Then came a rust killer; more washing and drying; followed by the sealer itself, which was a two-part system that required thorough mixing before pouring in the tank.

6.19 The tank had to be fully covered internally, so needed rolling slowly around for 20 minutes or so, until I was absolutely sure that all the internal faces had been coated. This also covered the inside of the cap, which was rusty, but might now prove reusable, if the air hole was still open. Excess sealant was poured into an old container and properly disposed of.

6.20 If the old cap had been salvagable, the very least that would be done would be to change the cork seal. Triumph caps were notoriously leaky, according to period road tests.

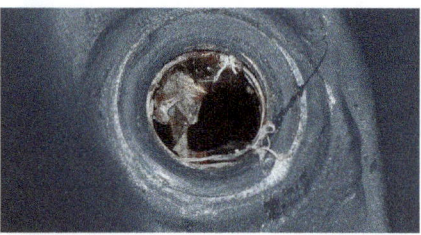

6.21 The sealant was left to harden in a warm environment, and, once set, the plugs were removed, which revealed some remnants that had to go.

6.22 I did not have a thread chaser in the correct size, but did have a tap, so used that to clean out the threads.

6.23 The tank had stuck to its mounting rubber on the way out. The old one was reusable, so received a smear of red grease to prevent future problems.

FUEL AND EXHAUST

CARBURETTORS

6.24 One thing I learned over many restorations was the misery that resulted from fuelling problems, often thanks to fitting remanufactured parts. Fortunately, Amal sell every part I needed to overhaul my bike's carbs, so quality was assured.

Amal Concentric Mk1

The Mk1 was a simple, reliable unit, with a couple of common and, sadly, serious defects, which are dealt with in the photo section. Replacements are inexpensive, and the new 'Premium' version addresses the most common issues at a slightly enhanced price. If you intend to retain your carbs after stripping and inspecting, Amal sell overhaul kits, with the required gaskets, etc, in one bag.

6.26 The number on the side indicates model and fitment, left or right. This right-hand version was used as the single unit fitted to a Tiger 750, early on in the production run.

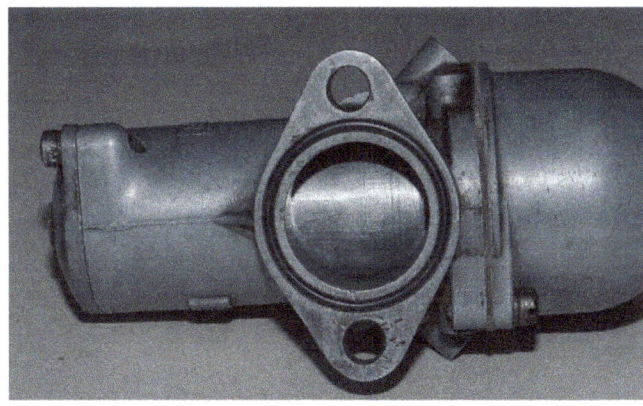

6.27 It bolted to the head, via this flange which relied on a rubber O-ring to seal the joint. It proved all too easy for owners to over-tighten, which distorted the flange and caused air leaks.

6.25 The much-maligned Amal Mk1 was actually a simple and reliable unit. New standard versions are available, along with an improved model, which addresses its known weaknesses.

6.28 The top was held by two screws. I was told that the threads in the body would often be found damaged.

HOW TO RESTORE TRIUMPH BONNEVILLE T140

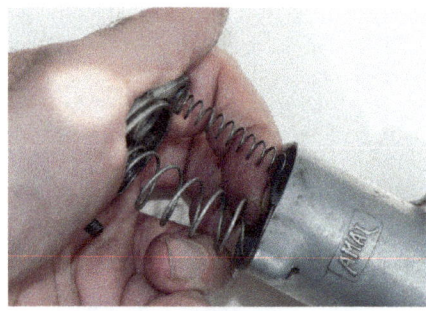

6.29 Once undone, the spring pressure pushed the top up and, once out of the way, the slide could be removed. Slightly fiddly, as the spring had to be compressed whilst unhooking the cable.

6.30 The spring should be 3in long. The air slide assembly next to it was also spring loaded.

6.31 I had a quick check of the body, which showed very light scuffing, but certainly nothing major.

6.32 It was the same story with the slide itself. This carb was in pretty good condition, so far.

6.33 The needle was secured by a clip, which was left alone for now.

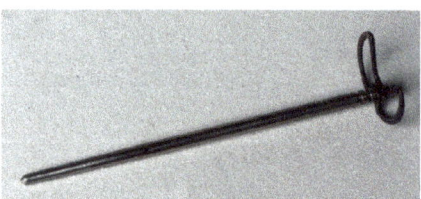

6.34 The needle itself showed as little wear as the slide and body. There was no ridging of the tip and the sides were still smooth. They can sometimes become pockmarked, but, again, this one was fine.

6.35 The carb bowl included a drain screw and the fuel inlet union. Like the carb top, two screws held the assembly in place. The carb was kept upright as the bowl was removed, otherwise the float assembly could fall out. All the fibre washers were replaced when rebuilding.

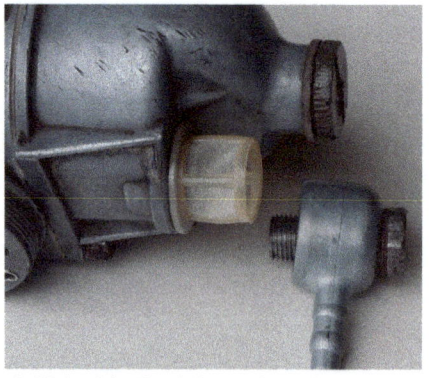

6.36 Inside the inlet union, there was a plastic gauze filter.

6.37 The float pivoted on a pin, which simply pushed out. The float needle valve loosely dangled off the bracket.

6.38 The bowl was thoroughly cleaned, and the float valve drillings blown through.

6.39 The float bowl gasket was automatically renewed.

FUEL AND EXHAUST

6.40 This float needle valve tip could have been checked for ridging. However, it would take such a small amount of wear to affect its performance, I decided that it would be best if it was renewed.

6.41 The jet stack was screwed into the main body.

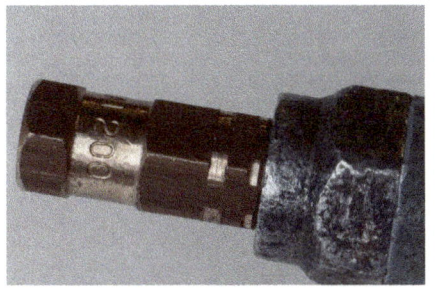

6.42 The jet sizes were clearly stamped.

6.43 The jet stack broken down: if they were to be reused, they would need thorough cleaning and blowing through. Needle jets should probably be changed, unless the mileage was known to be low, as it can be tricky to gauge wear as they tend to go oval. As everything else was in such good nick, I decided to leave them.

6.44 The throttle stop and mixture screws were removed. I counted the turns needed to get them out, so that they could be returned to their original position. However, if you do not know how the bike ran before stripping, that setting may be incorrect anyway.

6.45 The throttle stop screw was fitted with an O-ring, which was automatically replaced …

6.46 … as was the air screw. In addition, the sides at the end of the tip were checked for ridging.

6.47 The air screw hole was prone to blockages, so was given extra attention with compressed air.

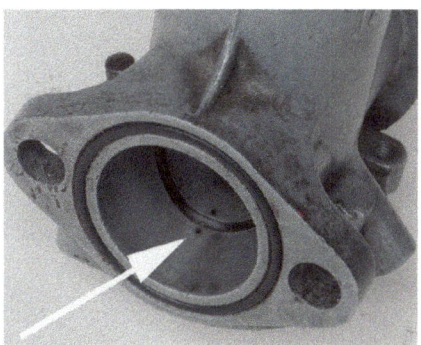

6.48 Check that air passes through all the relevant drillings in the carb.

101

HOW TO RESTORE TRIUMPH BONNEVILLE T140

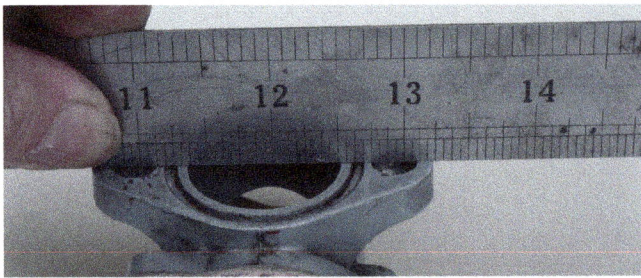

6.49 Distortion of the mounting flange is a common problem, due to over-tightening, as mentioned before. A straight edge was used to check that it was still level.

6.50 To double-check, some 400s paper was taped to a known flat surface, in this case a piece of automotive glass. An irregular, vaguely circular motion was used to ensure even coverage.

6.51 It came up fine and was obviously square.

6.52 Float bowls also suffer from distortion.

6.53 Exactly the same process was used to check that out.

6.54 It, too, was fine and suitable for reuse.

FUEL AND EXHAUST

6.55 This carb also had a tickler unit mounted on the side. This simply pushed the float down, to allow enrichment.

6.56 Reassembly was straightforward, I just made sure I knew where the needle clip sat in relation to the cut out.

6.57 The slide can only go back one way, as there is a guide notch and groove. Getting the cable back in with the spring compressed was just as fiddly as getting it out at the beginning.

Amal Concentric Mk2

These carbs arrived with the T140E like mine, to make sure the bikes met US emissions regulations. The slides on my carbs were stuck, which prevented removal of the tops, and even made splitting the cable junction box impossible, although that was undoubtedly hindered by an oversized ferrule on the single cable. So the carb bodies were inverted, and a capful of diesel poured into them, which was allowed to soak for a short time. This was sufficient to remove the old gummed-up fuel which was causing the sticking, and, once the slides were moving, the normal removal procedure could be followed.

6.58 My bike (and in theory all Es, but there were factory photos of that model in 1978 still sporting the old Mk1s, despite the Emissions tag) came fitted with Amal Mk2 carbs, an improvement on the previous model, but pretty similar in construction.

6.59 They were mounted to rubber inlet stubs with Jubilee clips, which helped reduce vibration-induced frothing, and banished the Mk1 problems with warped mounting flanges.

HOW TO RESTORE TRIUMPH BONNEVILLE T140

6.60 The slides on my bike were stuck, not uncommon on a project bike.

6.61 The carbs were inverted, and a small amount of diesel poured around the stuck slide and left for an hour to do its magic. On returning, a gentle push with a finger got them moving, which allowed the top caps to be unscrewed.

6.62 The carbs were joined at the back by a metal bar. If this bolt is loosened, the bar can be pivoted up out of the way. I removed it completely. The mounting hole on the other side should be oval, to allow some movement between the carbs to keep them level when refitted.

6.63 The front of the carbs were linked by the choke-operating bar, secured by pins and clips that levered off.

6.64 The fuel lines joined at the bottom of the bowl (as the Mk1 version), and were pulled off once their securing clips were undone.

FUEL AND EXHAUST

6.65 Splitting the throttle cable joiner was another irritant, as I mentioned in the strip down. Even with the carb tops off, it was reluctant, partly due to the inner circular cable clamp being stuck in the body. It was well-lubricated with silicone grease when I put it back.

6.68 A good soak in degreaser, followed by vigorous work with a firm toothbrush, worked wonders. There were still small areas in the corners that harboured some staining, but the finish was already looking pretty reasonable.

6.66 The inlet rubbers had rotted and split; again, a common problem on most restorations.

6.67 The bodies were dirty, but not horrific. It mainly seemed to be old fuel residue, rather than corrosion.

6.69 This carb had been leaking for some time, and old varnished fuel like this does need something stronger to shift it, so it was soaked in panel wipe for a short time, to loosen it all before the degreasing stage.

HOW TO RESTORE TRIUMPH BONNEVILLE T140

6.70 The choke mechanism on Mk2s was screwed into the body …

6.71 … and came out as a complete plunger unit. Individual parts can be bought, or the assembly is also sold as a complete kit.

6.72 The bowl had a large drain plug. All gaskets and seals were automatically replaced during the rebuild.

6.73 With the float bowl off, the internals were similar to the Mk1 version.

6.74 This needle valve did show ridging on the tip.

6.75 The seat where the needle tip sits included a small washer, which was very easy to overlook during the cleaning process.

6.76 The float was tested to check for leaks. Amal sells ethanol resistant 'Stay Up' versions as a direct replacement. However, I decided just to run the originals for a while, to see how things went. If there were any issues, changing them later would not be a big job.

FUEL AND EXHAUST

6.77 The bowl contains the pilot jet, which should be removed and cleaned.

6.78 The jet stack was accessible with the bowl off. It was covered by a mesh sleeve filter, which was not included in the parts list I had.

6.79 It just pulled off to reveal the jets themselves.

6.80 The main jet unscrewed as expected, although the size wasn't! The listing for my bike specified a 200 jet, and it had 190s.

6.81 The clue to the change was the American rear number plate on the bike, which was from Colorado, so the last owner probably lived high enough above sea level to have had to re-jet, to compensate for the thinner air.

6.82 The jet holder was then unscrewed to reveal the needle jet. Once again, unless certain of the age and mileage, it should be changed, as it would be very hard to accurately measure.

6.83 If everything was being reused, then a good clean, followed by a blow through with compressed air or carb cleaner, would be needed.

HOW TO RESTORE TRIUMPH BONNEVILLE T140

6.84 The body had to be checked for scoring, although the Mk2s suffer from this less as they are aluminium rather than the cheaper zinc-based alloy of the Mk1. This one was marked, but very lightly – more akin to polishing than gouging, so was fit for further use.

6.85 The slide, too, was an improvement on the earlier version, and coated for wear resistance. That coating was just starting to polish off in an isolated area, but, once again, nothing major, and certainly fine for several thousand more miles.

6.86 The pilot air screw and throttle stop screw were similar to the Mk1 and were removed, checked, their O-rings replaced, and the carb drillings blown through.

6.87 New rubber mounts were purchased, and the clips cleaned, ready to remount the carbs.

6.88 Although I had sealed the tank, I decided that, despite its ugliness, fitting a fuel filter would be a good idea, at least for the first few hundred miles. This one had an arrow for direction, so, although it looks upside down, it was in fact correct.

6.89 One popular option with owners is to junk the British carbs altogether and fit Japanese ones, mainly for reasons of build quality. I was happy to stick to my Amals, though.

6.90 Normally, the air filters would be paper elements inside alloy housings on each side of the frame. Mine had been junked in favour of cone foam filters. I tried to locate the correct boxes during the rebuild, but was unable to find any, apart from one set where the vendor wanted more than the price of a pair of tyres – a bullet I would have had to bite if I had been obsessed with a factory-fresh restoration. Boxes for the splayed-head Mk1 carbs were easier to track down.

FUEL AND EXHAUST

6.91 Removing the original filtration seemed to have been a popular mod, though – I saw a few with meshed bell-mouths like this, and others running retro pancake filters.

Bing

Triumph needed these carbs for its export bikes to the USA towards the end of production, as emissions regulations forced ever-tightening metering. Not many bikes were fitted with them, as production numbers had slumped dramatically by this time. The basic layout was similar to the previous Amals, but these were much more complicated carbs. They do suffer similar issues, though, such as slide wear, complicated by the addition of diaphragms. The finer drillings in the body means that cleaning them is tougher, and the idle jets and associated tubes are prone to clogging. The float needle and choke mechanism deteriorate with age, as well as the usual O-rings throughout. Some essential parts are available from specialists, as are complete carbs, direct from Amal.

CARB CLEANING
Manual

The old-fashioned way relies on lots of scrubbing, but, in many cases, it's perfectly adequate, and was the method I used. The carbs were stripped completely, then submerged in fresh degreaser for several hours, with regular sessions of agitation with a firm toothbrush. Many people recommend more potent cleaners, such as cellulose thinners or commercial carb cleaning dips, but these can be too aggressive for motorcycle carbs, and can damage the surface, leaving them pock-marked or with a permanent washed-out staining. The degreaser can be washed away with hot soapy water, but the carb must be thoroughly dried immediately, or the alloy will start to fur up in a very short time. All the internal airways and orifices were then blown through with compressed air. If that had not been an option, then carb cleaner could be used and aerosols often come with a fine tube attached, which would be ideal for this type of work.

Ultrasonic

This is the best cleaning method for carb internals, without a shadow of a doubt. Done professionally, though, it was not cheap: I was quoted around 70% of the price of a new carb to clean both of mine, plus another 20% if I wanted the outside aqua blasted. Chuck in the cost of a rebuild kit, and a set of brand new carbs looks very attractive. For the price of a basic professional clean, I could also have bought a budget ultrasonic tank off the internet, although these cheap units have a mixed reputation for both ability and longevity.

CARB FINISHING

Aqua blasting provides the best and quickest solution to brightening up old carbs, but it definitely isn't the cheapest. If your budget doesn't stretch that far, it is still possible to get a good finish manually. Discolouration and light surface deterioration can be sanded out with very fine, well-lubricated, wet-and-dry paper. Once clean, the main problem would be maintaining that finish. When manufactured, the carbs were dipped in a multi-chemical solution which etched the surface; that original finish degrades with age and cleaning, and is impossible to get back. Clear lacquer can be successful, but may alter the colour and eventually turn yellow and/or peel off; metal polish works in the short term, but needs constant reapplication to keep things looking good, which would be pretty tricky once the carbs are back on the bike.

FUEL AND ADDITIVES

Having spoken to lots of other owners, the general consensus is that these Triumphs run perfectly well on unleaded fuel, without the need for additives, or the expense of fitting hardened valve seats, which was good news. Premium fuels are, however, recommended for sustained high speed use, which was something I was unlikely to be indulging in. Ethanol was another issue, and much more of a concern. Internal tank rusting was seen as a real problem, as was fuel going off, when the bikes were stored, so additives and stabilisers seemed to be in common use for periods when off the road. Rubber seals in old taps can be attacked, but anything sourced today from a reputable seller, should be resistant – all of Amal's bits for example have been suitable for many years. These issues will become even more important, as ethanol levels increase in the coming years from the current 5% to 10%.

EXHAUSTS

Apart from some very early Vs, T140s came with push-in exhaust pipes, although very late versions reverted to a screw-in stub, as used at the beginning of production, with the pipes pushing over it. Push-ins can work loose, but a pipe spreader can often be used to take up the slack. If left loose, the head can be damaged, and repair will have to be left to a specialist, either to be built

HOW TO RESTORE TRIUMPH BONNEVILLE T140

back up, or tapped to accept the threaded screw-in stubs. The latter style is not without issues either: screw-in stubs, too, work loose and damage the threads, requiring professional attention. A push-in repair kit can be sourced from Triumph specialists, which uses a bolt-in securing ring to address the problem permanently.

Original-style pipes and silencers are readily available for the T140V and E, but other shorter-run models are less well catered for. As usual, quality comes at a price.

6.92 The chromed clamps at the head simply slid down, once the allen-headed bolts were undone. In my ignorance, I had assumed that they secured the pipes, but their main purpose seemed to be for heat dissipation through their large surface area. They were also decorative, of course.

6.93 The balance pipes were held in place by simple clamps, secured by nuts and bolts.

6.94 The down pipes were bolted to brackets secured by one of the engine case through bolts.

6.95 Push-in pipes can sometimes be a loose fit. An exhaust spreader could provide enough additional swell to take up the slack, but is not something most enthusiasts are going to have in their toolbox, so take your pipes to the local exhaust fitters and ask them to give them a tap; it would only take seconds.

6.96 Copies of the main models' silencers were readily available, as was a selection of sportier, non-genuine, two-into-one systems.

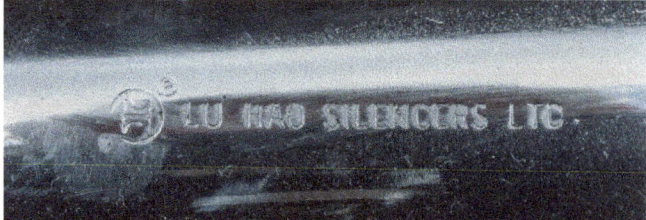

6.97 My silencers were in good condition, which saved some pennies. They were manufactured in China and appeared to be well made, with decent chrome. Had they not been usable, I think I would have selected this style in stainless steel, as they can be sourced from a UK manufacturer for a very reasonable price.

Chapter 7
Wheels

WHEEL REBUILDING
The wheels on my bike were in good condition, which was one of the factors which drew me to it in the first place. Rebuilding wheels is one area of the restoration where handing it over to a specialist would probably be a wise move, but the choice must be made carefully: a friend took his Moto Guzzi T3 to a local bike shop that knew 'a man in a shed' who could do it very reasonably. Once done, he was nearly pitched off, within a few miles of collecting it, as the wheel collapsed with several broken and bent spokes. Fortunately it all happened at low speed.

Once you have decided on your builder, there are many options: chrome, stainless or alloy rims, with stainless or galvanised spokes, or perhaps a re-chrome of the originals, but all come at a price. Rebuild both ends, with additional work cleaning the hubs, and the final bill could be the single biggest chunk of the running gear's restoration budget; if you have a Jubilee model, you also have to factor in the cost of the centre pin-striping, on top.

MAG WHEELS
Two types were fitted: Lester and Morris, depending on model. Either would need close examination for signs of cracking, and it is not uncommon for alloys to distort over time, as well, which results in excessive run out. If any of these issues apply, then replacement will be the only option, and finding replacements could prove difficult. If the wheels are sound, then refinishing would throw up the same choices as for the frame: powder coat or paint. If the latter, then base coat and lacquer provide a softer finish, but still with a decent shine.

7.1 The beauty of alloys definitely lies with the beholder. Restoring cosmetic issues is straightforward; anything more and replacement would be the likeliest option, assuming the correct ones can be found.

HOW TO RESTORE TRIUMPH BONNEVILLE T140

TYRES

The tyres on my bike were obviously hardly used, as they still bore their little moulding bumps and coloured stripes in the tread. A quick check of the DOT number revealed that they had been manufactured in December 2000, so were nearly 16 years old. The rear tyre was clearly marked for front use only, and was directional. Unfortunately, it had been installed the wrong way round, compounding the error. No matter – they were both destined for the dump, but that was not an easy process, as they were a nightmare to get off. The bead was stuck solid to the rim, and resisted all normal efforts to break it, so the wheel was hauled to the nearest tyre machine, and its air-operated bead breaker used. Even then things were not simple; the carcass was so hard that normal-length tyre levers wouldn't pull it over the rim, so long bars had to be used, with one person holding the rim whilst another levered away. It took a solid 20 minutes-per-tyre to lever them free. Once off, the inside of the rim was not bad, but the original rim tape had been replaced

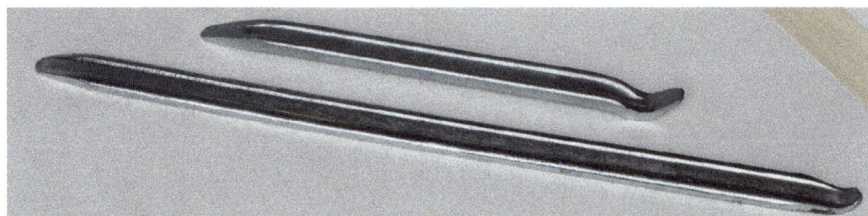

7.4 The tyres were incredibly hard, semi-fossilised, in fact, so standard irons were useless. A pair of extra length ones were bought, but it was still very much a struggle.

7.5 The rims were in good condition, as were the spokes, one of the reasons that imports from the USA remain popular, as wheel building isn't cheap. New rim tapes were fitted, as the old ones had rotted away.

7.2 The tyres on the bike had lots of tread and still had their coloured bands, so neither had done any mileage.

7.3 Inspecting the tyre sidewall gave the production date code – a handy check on any restoration project. They were more than a decade old.

7.6 New inner tubes were fitted, too. I have used Michelin ones for years; everyone seems to develop personal preferences, and this is one of mine.

7.7 New traditional treaded tyres were fitted for that classic look. Fortunately the actual compound was modern, so grip should be better than in the old days.

with Gaffer Tape, obviously at the last tyre change, as it looked like new.

The replacements were going to be TT100s, as the bike in the brochure picture sported them, and they had the classic look, combined with a modern compound, so the best of both worlds, as far as I am concerned. Unfortunately, they were out of stock everywhere I tried, and none were expected for several weeks, so rather than delay the build, I opted for a set of K70s – another quality option, with pedigree. A pair of new rim tapes seemed like a good idea, and some Michelin inner tubes – not the cheapest, but items I have used on all my bikes. Re-fitting was a struggle, despite using dedicated long tyre levers and lots of rubber lubricant, although the sub-zero temperatures at the time probably did little to help.

WHEEL BEARINGS

The Workshop Manual covered the process in a fairly short paragraph, which indicted that it must be straightforward, and so it proved. The bearings supplied now should all be sealed for life, which simplifies maintenance.

WHEELS

7.8 The bearings had obviously been done before, and the retaining ring removed and refitted with a punch. There is a special tool, but the manual suggests the use of the punch, in its absence.

7.9 Once given a couple of firm taps to get it moving, the ring unscrewed easily enough.

7.10 Spacers lifted out to gain access to the bearings underneath. It all came apart exactly as the book suggested it would.

7.11 The speedo drive ring also showed signs of having been on the receiving end of a punch. It can be removed using a decent-sized screwdriver in one of the slots and levering.

7.12 The bearings on my bike were sealed; replacements should also arrive that way, today.

7.13 Like this one, they provide better weather protection and remove the need to periodically strip and grease the bearings.

7.14 Bearing removal and replacement was by hammer and suitable drift, and was drama free.

SPEEDO DRIVE

The drive was taken from the rear wheel and was originally Smiths, fitted on the right-hand side. When the clocks moved to Veglia, it initially retained the right-hand drive box, before moving to the left, after a year or so. The change officially took place around late 1978, but there seems to be some considerable overlap: my '79 had the earlier drive fitted, but with Veglia clocks, for example. The drive ratios were the same, and the speedo heads were interchangeable, although the cables were a different length due to their relative positions.

7.15 Speedo drives differ, depending on the age of the bike, but all work in the same way. There isn't a lot you can do with them, other than cosmetic work externally.

7.16 The reverse side of the speedo drive should have the ratio clearly marked. A grease nipple was fitted, but it made sense to pre-pack under the lip of the cleaned unit, before it was fitted back on to the bike.

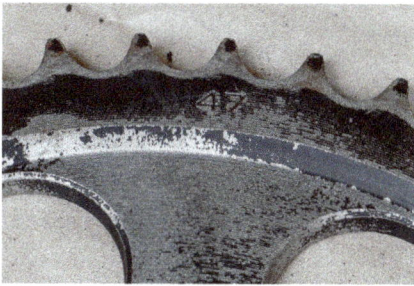

7.17 If original, like mine, the sprockets should have the number of teeth and the part number stamped into it. Given the likely use of T140s today, it might well be an idea to raise the gearing slightly for more relaxed cruising and improved economy – an option I would explore once the current set-up wore out.

Chapter 8
Paint

DECISION TIME

There was no doubt in my mind that no matter how good (or bad) the rest of my restoration efforts had been, most people were going to judge the whole project on the shine of its paintwork. Getting a good finish at home was perfectly possible, although it was unlikely to be show stopping, so it could have been time to consider handing things over to a professional outfit. If I had chosen to do that, then I would have had a look at its recent work on other bikes (or cars, if it was a general body shop), and made sure that I got a firm written quotation. I would have asked it to include 'final finishing,' which would ensure that any minor defects which I missed during the preparation stages would be its responsibility to spot and rectify. However, I chose to do the job myself. The most important attribute was going to be patience, as paintwork cannot be rushed, or I would just have ended up having to retrace my steps to rectify defects. The rewards, though, were huge: the satisfaction of seeing my bike shining in the sun, and knowing that it was all my own handiwork is hard to beat.

PAINT CHOICE

The options were very limited here, as home-application meant choosing 'safe' paints. Two-pack paint was totally out of the question, unless I could source an isocyanate-free version as the 'normal' stuff was potentially lethal, without the correct breathing equipment.

I intended replicating the factory colours, so my options were more limited, as it required using a base coat and a lacquer gloss top coat. There were only a couple of suppliers

8.1 This label wasn't kidding: do not be tempted by Two-pack paints under any circumstances. Even very short exposure can produce serious, long-lasting health problems.

PAINT

that could provide accurate colours ready-mixed off the shelf. The alternative was to take my tank to a local car paint supplier and let it mix some up, using mine as a sample. This can normally get pretty close, and, to be honest, colour consistency with all motorcycles of this era was pretty patchy, to say the least (Japanese included), so minor shade differences were not really a concern.

WHAT YOU NEED

Starting with the blindingly obvious: a spray gun. These varied enormously in price, but the good news is that, unless you invested in a powerful compressor, buying anything other than the most basic of guns is a waste of money. To be able to spray something even as small as a motorcycle petrol tank, a 2hp compressor, with at least a 50-litre air receiver, was going to be needed. Budget compressors are nearly all direct drive and tend to be very noisy in operation, so the neighbours were unlikely to impressed, even if they didn't notice the clouds of overspray that I was likely to create. Good lighting is essential when painting, so lots of fluorescent strip lights were needed, sealed, to prevent any chance of the flammable paint cloud being accidentally ignited. Some form of extraction system could be pretty handy, too, but I had to rely on leaving the door open.

Moving on to the personal requirements, the most important is a dedicated mask, with replaceable filters, suited to the paint type. I wore clean overalls, but a set of paper

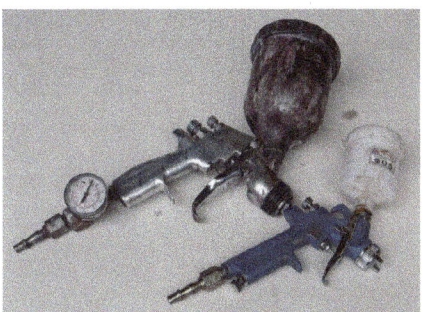

8.2 The smaller spray gun would be sufficient for motorcycle work, as small compressors may struggle to run anything bigger, for any length of time. The larger gun was about to go and get cleaned, by the way.

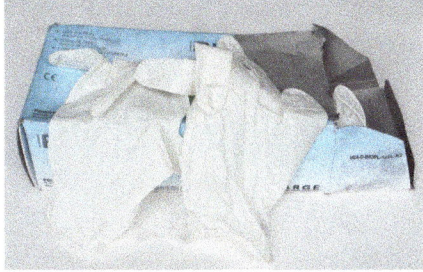

8.3 Disposable gloves saved my hands from contact with nasty chemicals, and kept them clean, too.

disposable ones, with a hood, would have been better. I used surgical gloves, which were bought from my paint shop, as not all of the ones on general sale can resist thinners and solvents, and they can rapidly dissolve into a gooey mess.

PAINT REMOVAL

The methods outlined in the Frame chapter were used, but an electric dual-action sander was a further useful addition for this job.

8.4 A dual-action sander may not be essential, but was handy, as it removed old paint quickly.

8.5 My bike's tank had stickers instead of badges, and they were damaged, so the old paint had to come off. A quick sand immediately revealed filler.

8.6 It became obvious that the area around the badge mounting holes had been knocked down, and filler used to level the surface, before the tank was painted black.

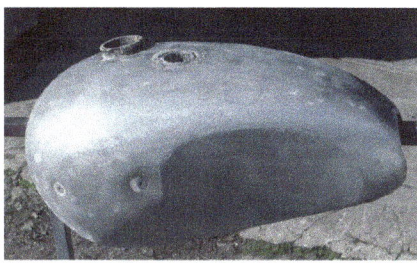

8.7 All the paint was sanded off. Tricky areas, where the sander could not reach, were treated to paint stripper.

ETCH

I read the instructions on the can of etch primer carefully before I used it: all versions contain acid, some included other unpleasant chemicals, so I followed the safety precautions to the letter. Aerosols are a possible option, here, as the amount you need for a bike was very small.

Applying etch by spray gun is ideal for practising technique. I set the gun up according to the instructions which came with it, aiming to get a spray pattern which was shaped like a thin upright rugby ball, with an even distribution of paint within it. I kept the gun roughly one hand-span away from the surface, and as parallel to it as possible, to follow the curves of the frame or panel exactly. I triggered the gun before I started my pass across the metal, and did not shut it off until I had passed over it on the other side. I tried to overlap each pass by 50%. There is lots of information online giving hints and advice on how to apply paint, but there was no substitute for practice, and the etch and primer coats were non-critical,

HOW TO RESTORE TRIUMPH BONNEVILLE T140

so they were a great way to build up confidence.

I automatically applied etch once anything went back to bare metal, as it allowed it to be moved around, if there were other issues, as with my tank, and it prevented corrosion. Once I was ready to paint after the repairs, I re-etched any bare metal once again.

8.8 Aerosol etch is the easiest solution for home use. Half a can will probably do two coats on most tank sizes.

8.9 The etch was applied once all the stripper remnants had been washed off and the tank dried.

FILLER

Despite having a poor reputation, thanks to misuse by the dodgier end of the motor trade, filler is an essential tool to the restorer. It was not designed to fill large dents, but, as a final thin skim to level surfaces, it's a godsend. It needs to be mixed thoroughly, to expel as much air before application as possible, and spread across the surface in long clean sweeps. Filler should really only be applied to bare metal. Modern formulations, though, stick to pretty much any sanded surface, so I wasn't too worried when a little etch was left behind as I prepared the surface. The filler was left to dry overnight, as, although, it may feel touch-dry quite quickly, it continues to shrink for some time.

8.10 I had discovered another very light skim of filler on the right-hand side of the tank when sanding, so the etch was now taken back, to reveal the low spots.

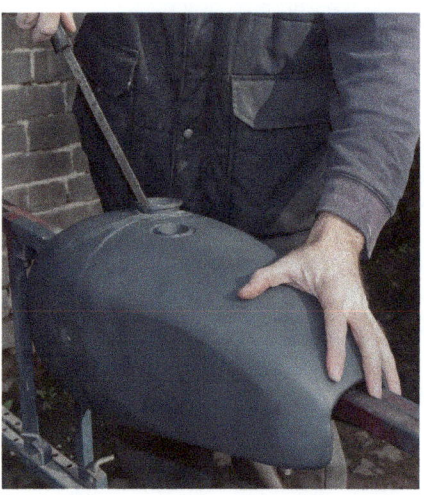

8.11 Before moving to the filler stage, the badge-mounting dents were pushed out, using a long pry bar.

8.12 All filler is not alike; I always go for an easy sand version, as cheap stuff can set like concrete. U-pol has been a long-time personal favourite.

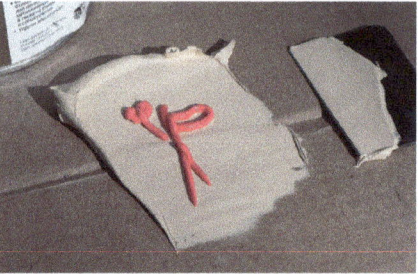

8.13 I mixed the filler, as per the instructions on the can: approximately a pea-to-golf ball ratio.

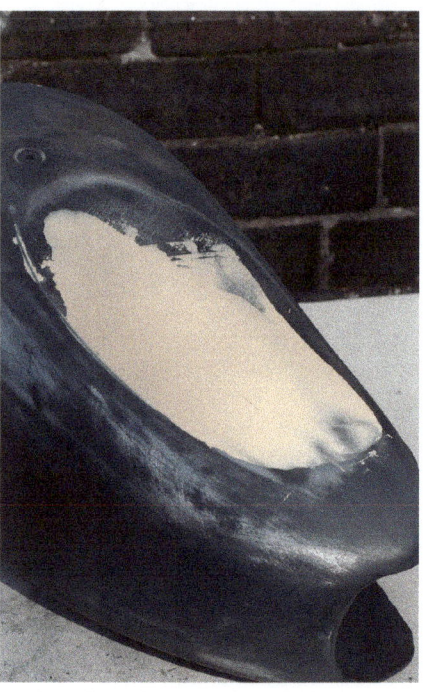

8.14 I made sure the whole area was covered. The smoother it was put on, the less work later sanding it back off, although I must admit I was rather lackadaisical about that, so ended up spending more time sanding than strictly necessary.

8.15 A hand block can help keep the surface flat, but may be of limited use depending on the curve of the tank where the damage lies.

PAINT

8.16 When sanding down filler, there will be a lot of dust. I protected my lungs with disposable face masks. I buy them from a car paint supplier: a large box costs no more than the price of a handful from a DIY chain.

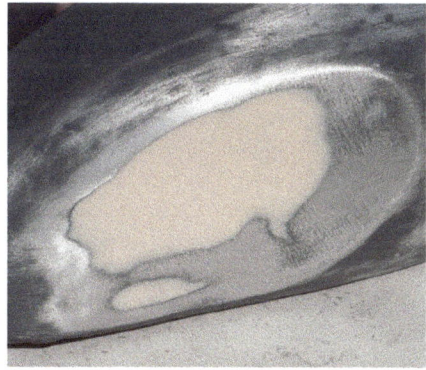

8.17 It was all taken back until everything was level. Unless you are very lucky, it might need a couple of goes to get it right. This wasn't bad, but still received another skim to bring the level up a little bit more.

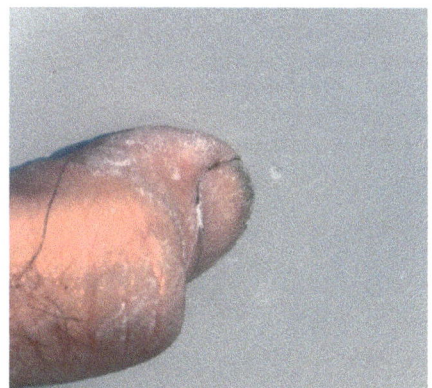

8.18 I ended up with a couple of small holes in the filler, as usual. Their presence can be minimised by thorough mixing to prevent air bubbles forming, but there always seems to be at least one.

8.19 The answer was stopper. This is a cellulose-based version, easy to apply and take off again.

8.20 I always add a lot over the defect, as I have found that this type of stopper can shrink back considerably. Once set, it can be sanded back to match the rest of the surface, leaving the hole plugged.

8.21 All the filler dust was blown away, then everything wiped down, followed by panel wipe to remove any oily impurities from handling.

8.22 The tank then received another coat of etch. The tank filler hole was plugged ready for the primer coat; a spray gun, with higher pressure and more paint coverage, was to be used, and I did not want paint getting inside.

WORKSHOP PREPARATION

Most home restorers like myself are not going to have the luxury of a dedicated spray booth, so some basic steps are needed to get the workshop ready. Cleanliness when painting is essential. Having gone through the filler stage, there was a lot of dust about in my workshop. I vacuumed everything accessible, then dampened the floor, blew off all surfaces, including the rafters and roof supports, and then left it all for half an hour. I swept out the shop, and repeated the process. If possible in future, I would have rigged up some sort of tent above the parts, to prevent any more debris which might be lurking from dropping during painting.

MASKING UP

I bought my masking tape from my paint supplier. It was far better quality than the stuff from general DIY stores. No matter what it was being applied to, I left a small folded section at the end to make removal easier later. I always removed the tape before the paint fully hardened, as there was a risk that it would lift the new surface during removal. The tape also tended to harden, once exposed to the chemicals in the paint.

8.23 For general masking, once again, make sure you buy your tape from a proper supplier; stuff sold for domestic decorating will not be tough enough.

PRIMER

If the etch coats had been very thick (they won't be, if you used an aerosol, as I did), I would have lightly sanded them, before I applied two coats of high-build primer. It is designed to fill minor imperfections, and can be flatted back to a level surface with 600s wet-and-dry. I used only clean water to lubricate the paper at this stage.

PRIME AGAIN

This final priming session was done with standard primer: a couple of coats was enough. I flatted off, once more, with 800s wet-and-dry, then throughly cleaned

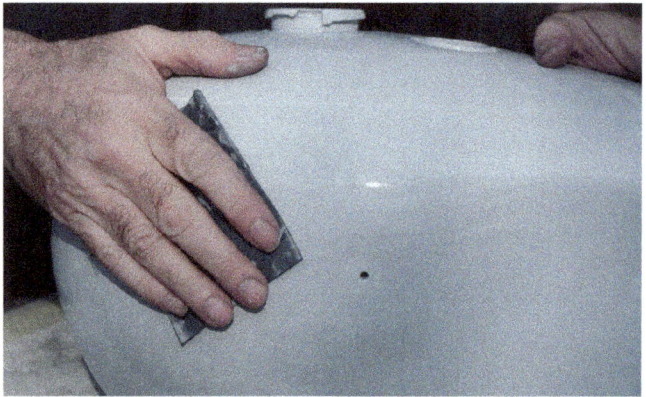

8.24 Two coats of primer filler were blown on, then, once dry, wet-flatted back.

8.25 The residue was removed using this blue paper. It is very strong, even when wet. I bought it ... you guessed it, from a proper paint supplier.

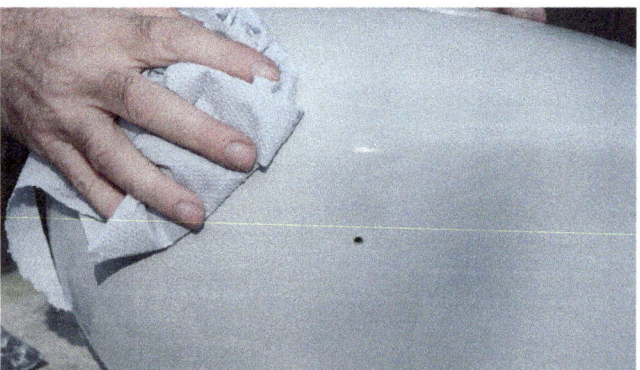

8.26 More primer, more flatting, more residue that needed to be wiped off, followed by more panel wipe, then a tack rag, so the surface was ready for the colour coats.

the surface. I wiped away any residue, then ran over the surface with a tack rag, which got rid of any final small particles of dust. The surface was then ready for the colour coats.

BASE COATS

The most common base coats available from most small paint suppliers are two-pack versions. Fortunately, unlike two-pack gloss, the base coats don't require the use of isocyanate hardeners, and are therefore no more dangerous than cellulose, or other paints. I applied three coats. The finish, which was matt, had only to be uniform and free from runs or other defects. If there had been any, I would have flatted back the surface, to lose them, and reapplied the paint once fully dry.

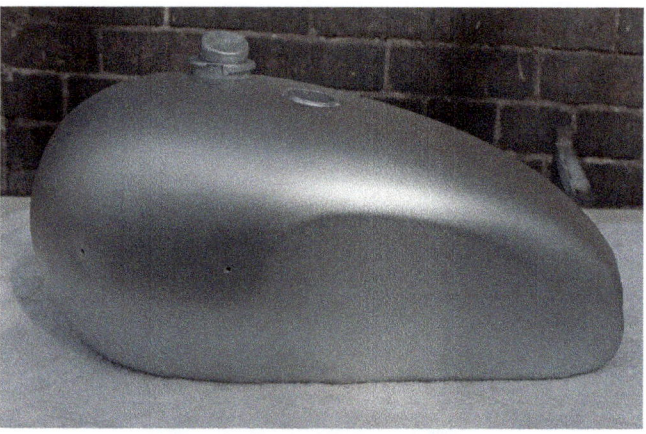

8.27 The candy red had to go over a silver metallic base coat, which was applied next.

TWO-TONE

Triumph produced some lovely tank detailing, with split colours and sweeping lines, all done by highly-skilled painters, not simply by applying transfers. The tricky part was trying to replicate that, especially when I discovered that the factory used preformed solid masks to speed up paint application. The only alternative, at home, was slow and careful masking. The second problem with trying to copy Triumph's work was that the lining between the two colours was done by hand, and sometimes there was a double stripe to contend with, which would really stretch my abilities, if it was going to look okay.

8.28 Once dry, the tank had to be prepared for its split colours, which meant masking off. Normal tape is of no use for this, as it allows bleed-through at the edges, so fine line tape was required. Its relative size can be seen here, as it rests on normal tape.

PAINT

8.29 It is stretchy and flexible, so moulded itself to the desired shape, which was laid out freehand. Paper templates would have been another option.

8.30 The tank badges were replaced temporarily, to make sure that the proposed lines would sit correctly.

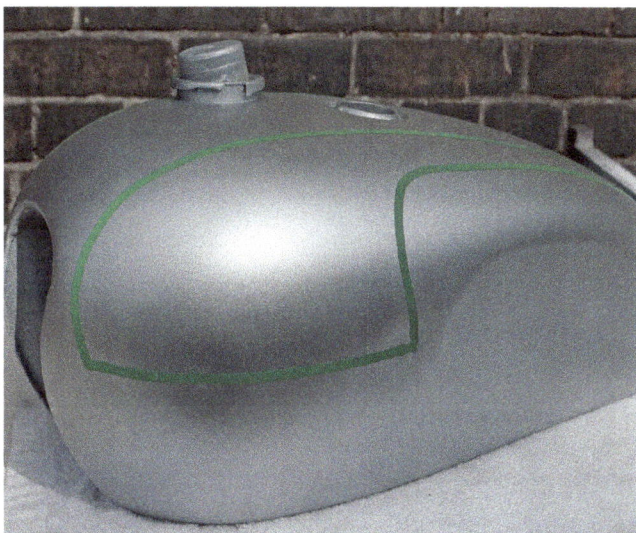

8.31 The bottom edge was just left straight, as it would be hidden by the badge.

8.32 The section that would be remaining in silver was then masked.

HOW TO RESTORE TRIUMPH BONNEVILLE T140

8.33 Factory colours changed with each model year, but the brochures did not always accurately reflect those changes. The 1979 publication showed a silver and candy red export model with gold stripes, which was the look I was now aiming for, although that combination was never put on sale that year.

CANDY

My bike's paintwork was non-standard black, and the American Title (log book/V5) did not have the original colour listed. In 1979, the factory churned out bikes for the USA in Candy Apple red and black, silver and black and, finally, blue and black. A glance at the brochure for that model year, though, clearly showed an export model with a red and silver tank. The pin striping was meant to be in black for that year's models, although the red/black combo definitely carried gold pinstripes in the photo, so there were obviously some variations. I decided on red and silver, with gold striping, on the basis that Triumph must have done at least one like that, if only for the photo. Applying Candy was tricky, as it was almost translucent, and had to be built up with lots of coats. Coverage and gun pressure affected the final colour, so I decided that it was best to do all parts at the same time, to make sure everything got an equal amount.

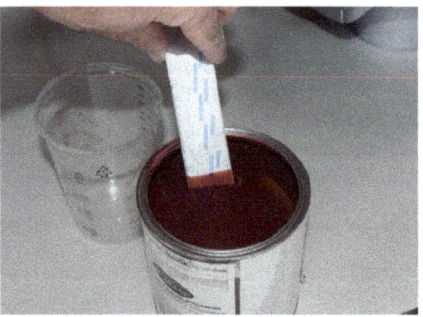

8.35 All the paint needed to be thoroughly mixed, as it tends to settle out in the can when stored. Filtering the mixed paint into the gun also helps prevent unwanted crusty contaminants getting in.

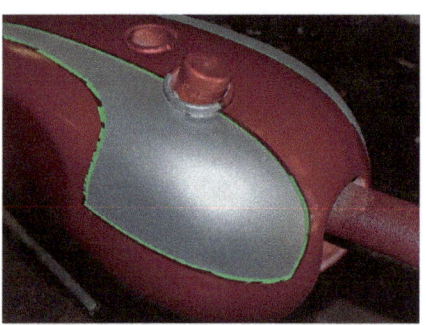

8.36 The chosen candy base coat was semi-translucent, and the colour only started to build up after many coats. I think I was up to around a dozen before I started to think the colour was getting there. Once happy, the masking tape was carefully removed whilst the paint was still slightly tacky. It was a great feeling, when the silver resurfaced.

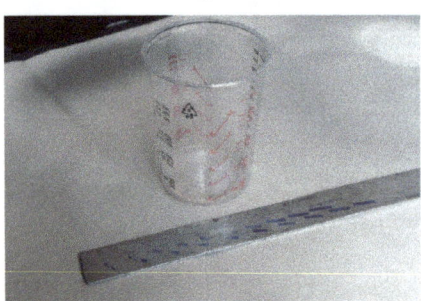

8.34 The paint needed to be carefully measured out in relation to its thinners. Plastic measuring cups are ideal, and provide a clean receptacle in which to mix it all, as well. The measuring stick next to the cup would have been another option.

8.37 The fine line tape left a clean sharp line between the colours, which would not have been possible with ordinary masking tape.

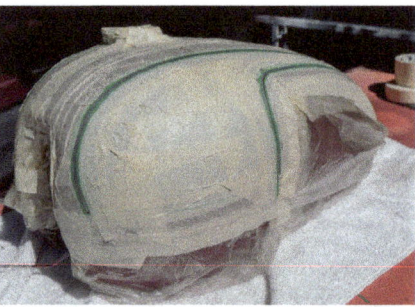

8.38 As my hands were not steady enough to attempt doing the pin-striping free style, it was back to more tape and masking, once the candy had dried. This time, I used a combination of tape and plastic masking sheet to cover up the areas to be protected.

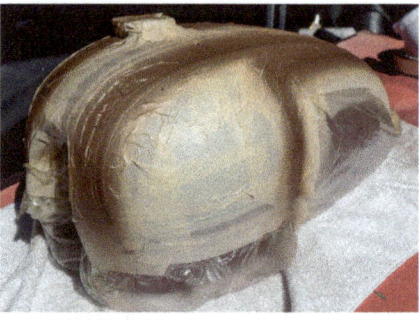

8.39 Gold base coat was then applied, and the tape removed, once again, before it had fully hardened, and still retained some flexibility.

LACQUER

This is what would give my bike its high gloss shine. Single pack lacquer would do the job, but can fade quickly; two-pack is far better, but you *must* get the non-isocyanate version, as mentioned above. It is thicker than its more dangerous cousin, and would require more post-spraying work to get the best finish. Alternatively, I could have handed my panels over to a pro shop for this final critical stage.

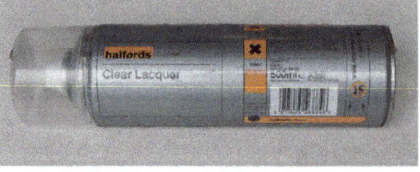

8.40 The clear lacquer then went on, 1K in the gun as a home user, but an aerosol remained an option. I have used this one before for small parts and it gave a decent finish.

PAINT

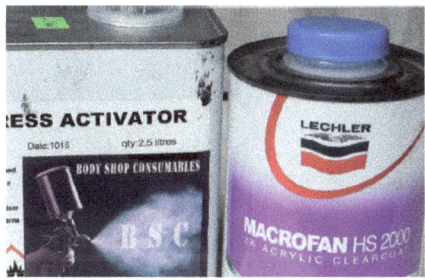

8.41 2K gives the best and longest-lasting shine. **Note: it is lethal for home use**, but, having got the tank and panels to this stage, you could take them to a pro paint shop to apply 2K for you.

FINISHING TOUCHES

Having left the paint surface alone for a couple of days, it needed to be closely inspected. It had dulled down very slightly, and there were minute specks of dust trapped in the surface, which I had expected, given the workshop in which it was applied. I used 1500s wet-and-dry paper, very wet and cut with a little plain hand soap, to gently flat the surface back to a uniform dullness. Fine brazing paste was applied to a small area, and polished with a buffing machine. I already had a buffing machine, but, if not, I would have borrowed or rented one, as, although it was perfectly possible to do it all by hand, I knew just how arm-achingly tedious it is. I made sure not to flat off the paint close to panel edges, or places where the machine was going to be a struggle to get at – I hand-buffed the original surface in these instances. Once the shine had been regained, I applied a good quality polish with soft, clean cloths.

8.42 Once the lacquer was fully hardened (after several days), I checked for any imperfections, and found some dust in the top layer. It could be removed by first sanding gently back with wet-and-dry, or individual bits rubbed out using a small de-nibbing pad.

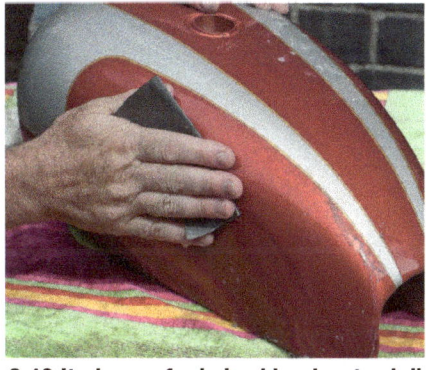

8.43 It always feels bad having to dull down the shiny coat that you worked so hard to achieve when putting the lacquer on. I used 1500s wet-and-dry, with loads of water, to flat out the dust.

8.44 It was slow going with such fine paper, but eventually I got back to a level flat surface free of defects. At this stage, it was hard to believe I would ever get the shine back.

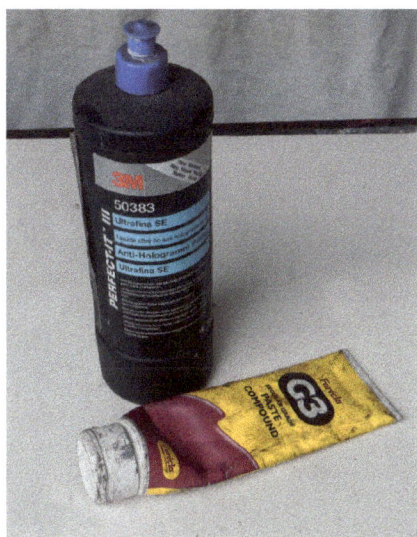

8.45 Brazing paste would now do the work for me. It comes in different grades, to suit the amount of abrasion needed to bring the surface back into shape.

8.46 A machine buffer made light work of this tedious job; doing it by hand would take some considerable time.

8.47 The mop head had been well used before, so it was thoroughly washed to make sure that no impurities would be transferred to my new paint. The soap was then washed off with clean water, as it was also possible for its residue to mark the surface.

8.48 It was a nerve-wracking experience, as the buffing process heated the surface as I worked, so burning through was a potential problem. I never fail to be amazed, though, as I watch the shine slowly resurface, and, after a few minutes, it was glossy once more, as the reflection shows.

8.49 Lots of wax finished the job, sealing in that shine.

HOW TO RESTORE TRIUMPH BONNEVILLE T140

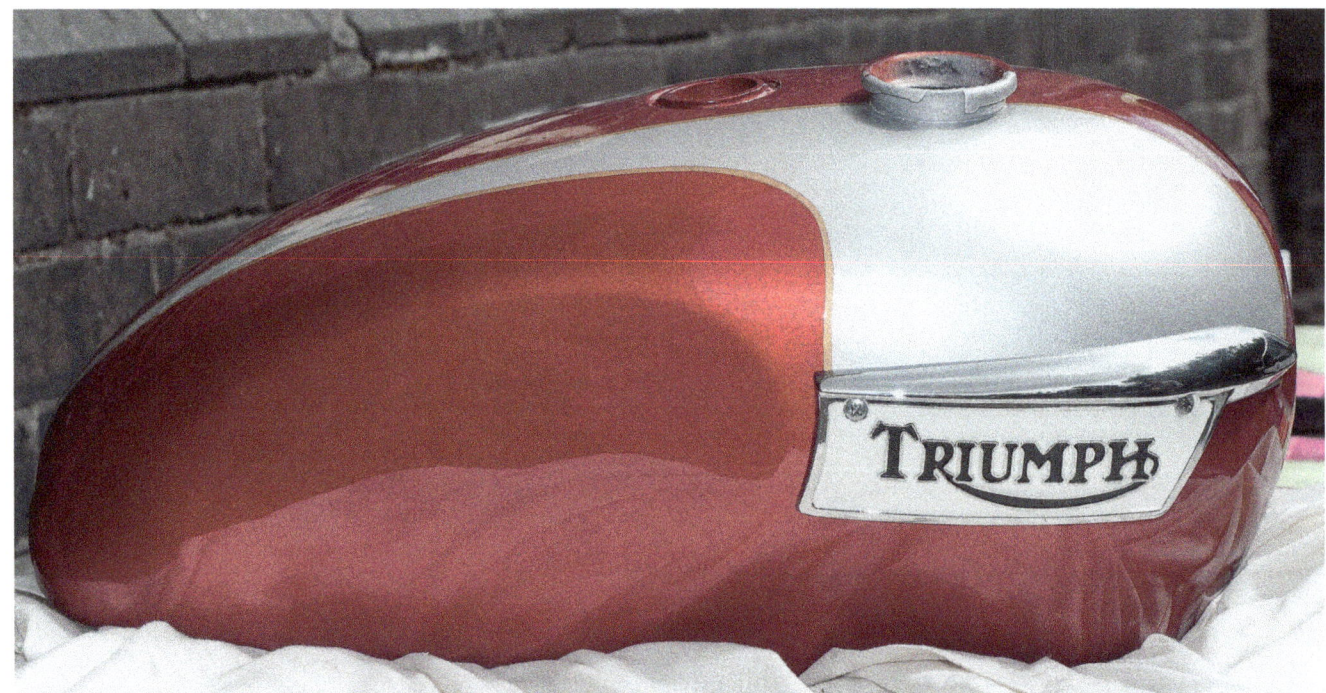

8.50 The finished tank: not up to Meriden's standards, but very presentable, and completed for around one tenth of the cost of farming it out. The process was also hugely enjoyable, if a little tense at times. I also really liked my choice of colour combination, original for the year or not.

PAINTING FIBREGLASS

This was used on side panels on later bikes like mine, and was dealt with in the same way as the tank. I panel wiped, then lightly flatted with 240s production paper, as I did not want to break through the gel coat. Wet flatting wasn't a good idea, as the surface was slightly porous and so might trap microscopic amounts of water which would ruin my top coats. I avoided etch primers, too, as they could react; I just used normal primer, and then applied the top coats.

8.52 Silver base was sprayed on, but, in another example of the perils of rushing, I had forgotten that there were a couple of minor imperfections that I had wanted to address first.

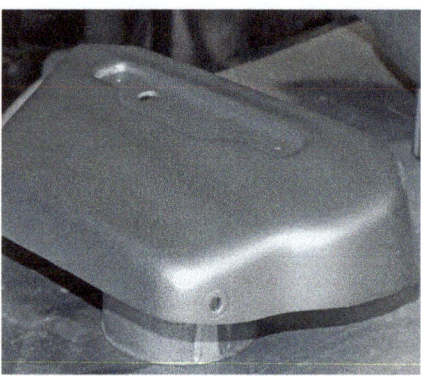

8.54 Once flat, I reapplied the silver base. Several hours had been lost needlessly through my impatience.

8.51 The fibreglass side panels were dry sanded until they were uniformly matt. They may have been moulded in that colour, originally, but they had been painted, and underneath were signs of Candy Red. My paint choice had been vindicated.

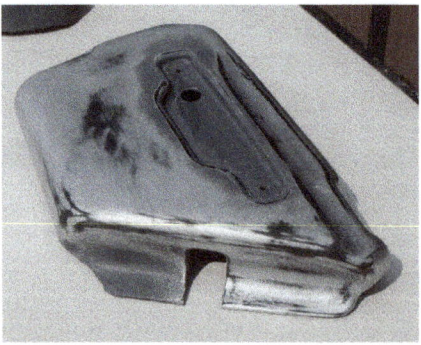

8.53 I took the opportunity to take it all back down again, to make sure that the surface was completely flat, then filled the imperfections.

8.55 As with the tank, the candy took an age to develop any depth of colour ...

PAINT

8.56 ... but I got there in the end. It was ready for lacquer.

8.57 Once lacquered, the panels were flatted back, just like the tank, as they had ended up with even more airborne contaminants in the surface.

8.58 The switches were a bit scruffy and faded.

8.59 They had lost paint in several places, a common fault on older bikes.

8.60 They were flatted back to a level surface, etched, then primed and painted in black.

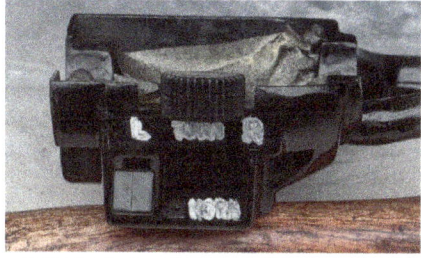

8.61 I tried painting just the letters with a small brush first of all, but the grooves were very shallow. I then tried covering the whole area, and allowing the paint to go slightly tacky, before wiping the excess away with panel wipe.

8.62 It worked, but the colour wasn't strong enough, so they required a second coat.

8.63 They turned out to be acceptable, but far from perfect. I could have spent more time trying further coats if I had felt the need. For a show bike, replacing the switches would be a better bet.

123

Chapter 9
Trim and brightwork

TRANSFERS/DECALS

There were several stickers scattered around the T140 in a rather haphazard fashion – just ask a few people exactly where the 'Made in England' sticker should go, for example, and you would quickly realise that things were put in place differently each day on the production line. Full sets of the correct stickers are available that cover most model years, plus there were some water slide versions around, when I was looking, as well. Stickers are simple to apply: just peel from the backing paper, position and apply. A gentle rub over, working from the centre out, should see them firmly and cleanly attached. For the slide type, drop them in water for a minute and the paper will curl up, remove, and leave it for another minute, and it will start to level out again. At this point, start to slide the transfer off the backing paper. Once there is a small area free, put it in place, then from that edge, slowly slide the backing paper out of the way. These slide decals are fragile, so it would be advisable to apply a little very-lightly-soaped water to the metalwork first. When placing the transfer, this allows some

9.1 I took photographs of all the sticker types and locations, as information on their exact position was lacking and is the subject of much debate.

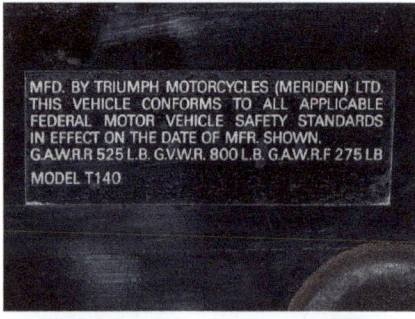

9.2 Most were stickers like this, rather than transfers. They are a lot easier to place, but always retain a visible line at their edge if examined closely, even if lacquered.

9.3 Old stickers, such as this, were another moot point. Should I have removed them for a factory-looking rebuild, or welcomed them as patina, part of the history of the bike?

9.4 Replacements for all models are around, and the ones I bought seemed very good copies.

TRIM AND BRIGHTWORK

9.5 I received several opinions on where the 'Made in England' sticker should sit, with some ideas differing by very small amounts indeed. I have a feeling that they were stuck on with less-than-absolute precision at Meriden, so I did not worry too much.

9.6 Small details, such as these fork-top labels, can make a huge difference to the finished look of the bike.

movement to get things lined up without undue strain. Once in place, excess moisture could be gently squeezed or dabbed away with a cloth. These transfers are designed to go over base coat paint, then sealed by clear coat lacquer to finish. Stickers might not be suitable for lacquering, so check with the vendor.

SEAT REPLACEMENT/ RECOVERING

My seat was the correct 'dropside' style, and in pretty good condition, apart from a small tear in the cover, which sadly meant replacement time, as I have yet to come across a decent way to make small repairs.

If a replacement seat seems easier, then all types are available, and, in theory, all should be interchangeable, with the only proviso that you have to choose between

9.7 The seat did not look too bad at first glance, and probably fine for a day-to-day ride.

9.8 The Triumph markings had faded badly, though …

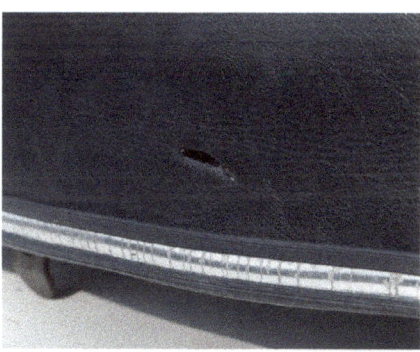

9.9 … and there was a cut in the side, which would not really be repairable, so a new cover was needed.

9.10 I made a note of the sticker position before peeling it off.

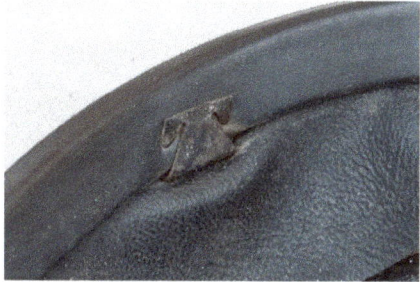

9.11 The seat cover and bottom trim were both held by metal clips that were tricky to remove. They needed to be levered away from the cover to free the teeth inside, before being pulled away.

9.12 At the front, there were fold-over metal tags pushed through the cover. They had to be straightened out.

9.13 The seat post was rusty, so was unbolted for attention later.

HOW TO RESTORE TRIUMPH BONNEVILLE T140

9.14 One pair of hinge bolts was brand new and came out easily.

9.15 The others were seized solid. I decided to leave them until the cover was off, when I would have access to the other side.

9.16 The seat base rubbers were levered out with a screwdriver.

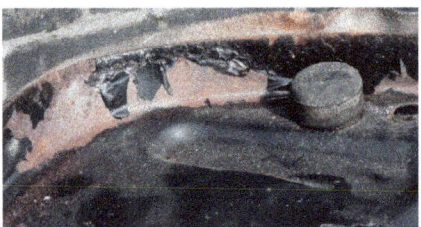

9.17 The powder coating on the base had gone the same way as the stuff on the frame, which had led to rusty patches.

9.18 With the cover and foam removed, I was able to access the back of the seized bolts, and administer some releasing oil.

9.19 The seat base top was rusty, but not heavily pockmarked.

9.20 The underside was slightly worse, but at least it would not need welding, as was so often the case on previous restorations.

9.21 The edge of the seat pan was dented in one place. The dent was removed using a hammer and dolly.

9.22 The foam was in remarkably good condition: they often turn dusty and crumble, in which case replacement would have been the only option.

TRIM AND BRIGHTWORK

9.23 I used paint stripper on the old base, as the curved shape made accessing with a sander tricky. It did not take long to work.

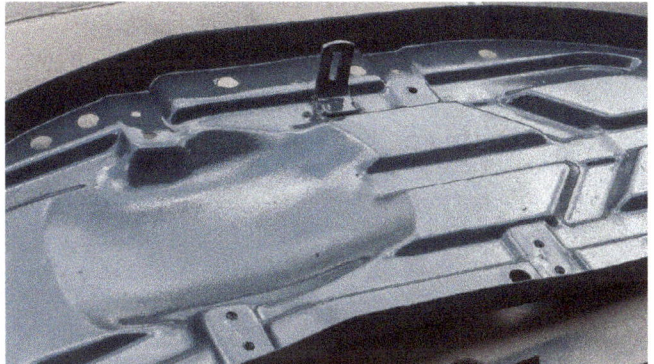

9.24 The base was then cleaned and etched. When fully dry, it was given a coat of gloss black paint.

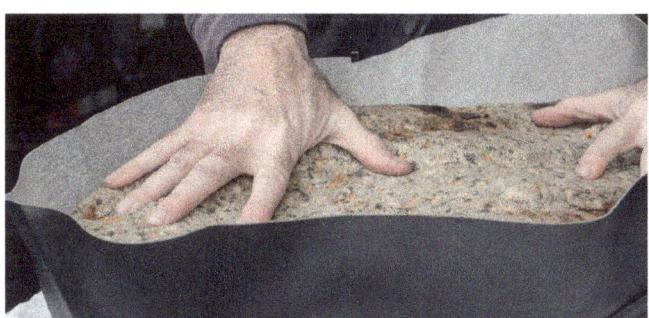

9.25 The new seat cover was placed on a clean cloth, and the foam inserted into roughly the correct place.

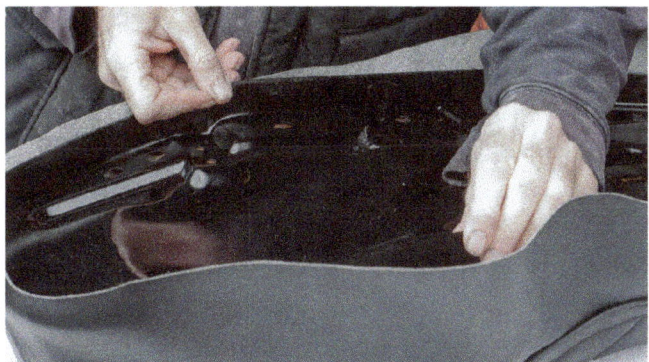

9.26 The metal base was then laid on top, and more minor adjustments made, until I was happy that it was all central.

9.27 The cover came with new clips, but not as many as originally fitted, so I rescued a couple of the old ones, cleaned them and painted them black.

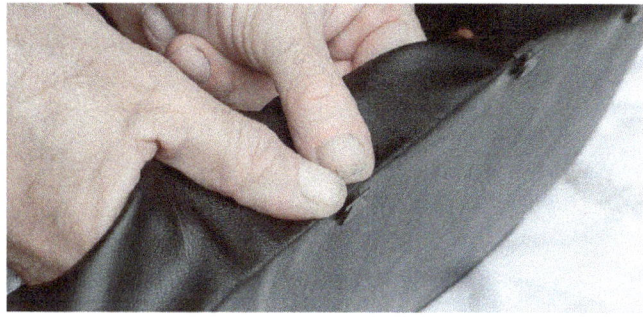

9.28 The seat had to have some pressure exerted on it, to keep the cover taut as it was fitted. I found that it was better to push on the securing clips only partway, until a complete circuit of the cover had been made. Had it needed to be tighter, they could then have been moved without ripping the new cover.

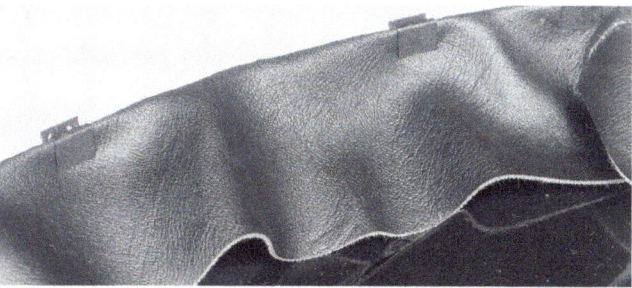

9.29 They were tapped down fully, once the cover was sufficiently tight.

9.30 The old edge trim had cracked, as it was made from plastic. The new cover came with a length to replace it.

HOW TO RESTORE TRIUMPH BONNEVILLE T140

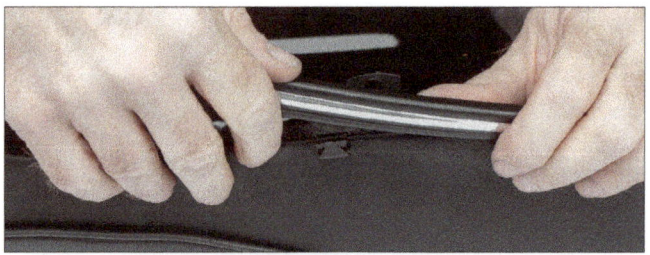

9.31 It simply pushed down over the seat edge to be gripped by the clips.

9.36 The seat catch now needed some attention.

9.32 The excess material on the underside was trimmed back with scissors.

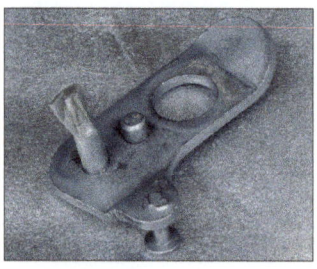

9.37 Zinc plating would be the best option on small items like this, but impatience made me just clean it and paint it.

9.33 Aerosol carpet adhesive was perfect for sticking down the loose edge.

9.38 The rusty hinges were stripped, etched and painted.

9.34 It made for a tidy job, but excess glue had to be wiped away with panel wipe, which lightly marked the paint, so some minor touching up with a brush was needed.

9.39 The seat rubbers looked grim, but closer inspection revealed that they were not actually damaged.

9.40 They were cleaned with panel wipe, then given a light rub with silicone grease, and came up perfectly well.

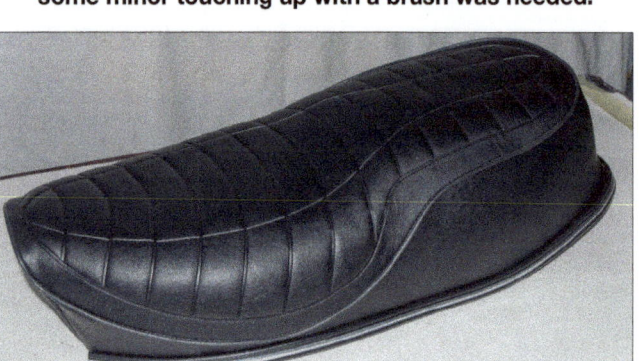

9.35 The finished seat looked plump with no wrinkles, which was the aim.

9.41 The seat locating pin assembly was very simple. It, too, was cleaned for re-use.

TRIM AND BRIGHTWORK

9.42 The seat buffers were lightly lubricated with silicone, to help get them back in easily.

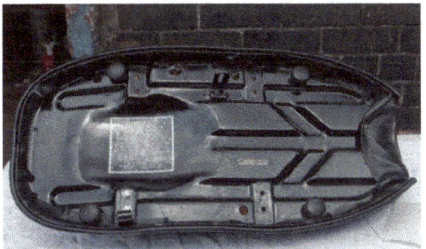

9.43 The under-seat stickers were applied in approximately the same place as the original.

9.44 And the rear hinge bolted back. The front would be done once the seat was threaded back into place.

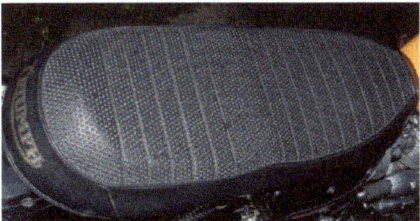

9.45 Ultimately, the seat can be swapped for one from any era you like, as long you stick to US and UK shape. I am a great admirer of this early basket-weave top, which could be retro-fitted to my E model without too much effort.

UK or Export, as they had slightly different shapes at the front. I was tempted to change, as I prefer the look of the earlier seat with the basket-weave top, and spoke to a supplier who said it was possible, and that, by adding a late model 1982 seat hinge and cover, I could also lose the seat check strap, as a stop is built into the hinge. Should you have an early bike and fancy a late seat, then that is also possible, as long as you add the seat catch plate from the later model. In the end, I decided to keep the bike looking original, but the option was there.

INSTRUMENTS

Restoring these bikes' instruments may be a problem. Initially fitted with Smiths units, these were phased out, starting in late 1978, when intermittent supplies forced a change to Veglia. Copies of the latter can easily be found, but the earlier clocks would probably have to go to a specialist restorer. Electronic modern versions are made, but, I was told, only in negative earth, so check first. The main issue, though, is the variety. Initially marked 'NVT,' those were obviously dropped when the Co-op took over and applied the Meriden mark, although I was informed that some were blank in this period. The change to Veglia does not help the restorer either, as some were marked 'Triumph,' others 'Veglia' and, again, some allegedly remained unmarked, so, if replacements are needed, getting the correct ones could be problematic.

The instrument panel fitted to '79-onwards bikes had a centre console, with the warning lights, which had migrated from the headlamp where they had resided on earlier models. The clock holders and console bottom came off together, and were secured by two 9/16 bolts, which pass through rubber bushes before locating in the top yoke. Mine appeared to have been powder coated, and had received a much thicker coating than the frame, so stripping took a bit longer. Once all the old stuff was removed, it was etched, and then given a top coat in black to finish. Pre-E models had individual clock holders made from chromed steel; copies in stainless steel are around.

9.46 The instrument layout was pretty similar on all models – but for the arrival of the central console with the E model.

9.47 The earlier Smiths instruments were in individual supports …

9.48 … with the light switch and tell-tales mounted in the headlamp itself.

9.49 The ignition switch on the earlier bikes was mounted in the left-hand headlamp ear.

HOW TO RESTORE TRIUMPH BONNEVILLE T140

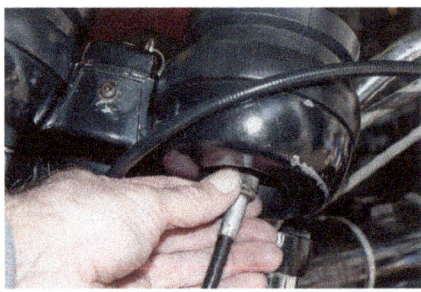

9.50 The speedo and tacho simply lifted out of their anti-vibration rubber mounts, once their respective cables were released.

9.54 The others were really tight and needed pliers to wiggle them loose.

9.57 The alloy instrument panel was secured by two bolts passing through rubber sleeves.

9.51 The bulb holders pushed through a rubber grommet which secured them.

9.58 It was chipped and sun-bleached at the front.

9.55 The bulb sleeves were pushed out, and the panel rubbed down and painted.

9.59 The old surface was stripped, etched and painted like the switch panel.

9.52 The switch and warning light panel was held in by screws into speed clips.

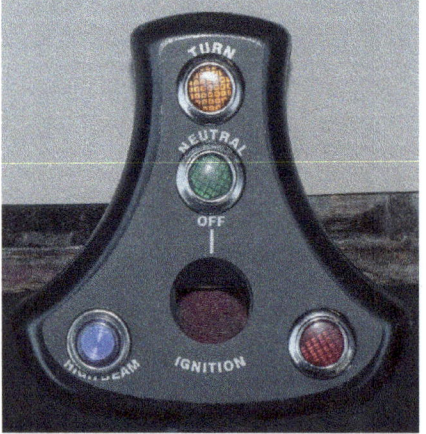

9.56 It received a new sticker, which wasn't printed properly, as the 'High Beam' lettering was partially obscured, and 'Oil' was almost completely obliterated by the bezel. A new high beam warning light was needed, and, when it arrived, the lens wasn't the same as the originals. Not important on my bike, maybe more so on a more meticulous restoration.

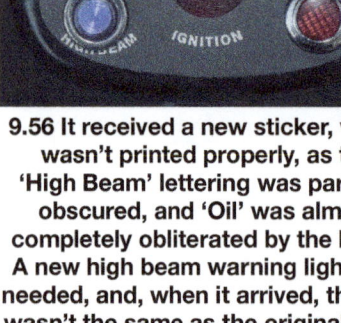

9.60 The last job with the clocks was to polish the lenses. If they had been badly hazed, I would have used very fine brazing paste to clean them.

CHROME PARTS
This is another area where the rebuild budget could be put under considerable strain. Replacement is often the cheaper option, but as we have seen previously, the quality of aftermarket parts is often very poor, so it may be better to bite the bullet, and have the original stuff re-chromed, if possible.

9.53 Once they were undone, it pulled up, with a couple of the warning light bulb holders left behind.

TRIM AND BRIGHTWORK

Mudguards

Most T140s had chrome guards, until 1981 when stainless versions were introduced. In 1977, the Silver Jubilee had painted guards, and, the following year, the two-tone brown models had them, too, as did the later Executive and TSX models.

9.61 The front mudguard had a small crease where the chrome had also flaked off. It could have been knocked out, but I suspected that the metal had probably stretched a bit. The price of a re-chrome after rectification was not far short of a new mudguard …

My bike had a small dent and scratch in the front guard, which I had failed to spot when buying it. I had to decide between straightening it and re-chroming, or replacement. I had seen some stainless mudguards, which were tempting, but when I tried to order, they were out of stock everywhere, and no one seemed to know when they would be back. The aftermarket chromed versions all seemed to be for the later bikes, which was fine for me, but, if you have a rear drum model, you might be stuck with the disc type rear guard, with the cut out. The rear mudguard on mine was held by a small bracket at the front, and through bolts on the frame loop at the rear, all fittings were $7/16$, as were the bolts securing the number plate support bracket, at the rear of the mudguard. Pre-1979, there was a separate metal bridge piece, between the rear shock absorber top mounts. All the fittings were rusty, so bikes subjected to European weather would almost certainly require new ones.

9.62 … which was my choice in the end. UK-made, it fitted well and looked like the original. I kept the old guard to try beating out the damage later.

9.64 The new guard was only a blade, so the old bridge piece had to be unbolted. These nuts were okay, but I suspect that would not be the case on a bike that had spent years in the UK.

9.63 Earlier bikes had a different stay layout.

9.65 Once off, the bridge seemed to have received plating to the lower exposed section only. Fortunately, it wasn't rusty; if it had been, new replacements are available.

9.66 The underside of the blade and the bridge were covered in Waxoyl, and I made sure it was well-brushed into the rolled edge. It would wash away over time, but re-treating it each year, before winter sets in, would not be too much of a hardship.

9.67 The older chrome of the mudguard bridge and the wheel rims were cleaned with an old favourite, Solvol, and it all came up a treat. For the new chrome, I used a gentler cream formulation out of a bottle.

9.68 Once mounted, the new mudguard looked great, but was misaligned at the front. It was fine at the rear, so was slightly twisted. Gentle pressure was applied to straighten it.

Headlamp/fork shrouds

Replacement headlamp bowls can be found for all types but some I looked at had very poor chrome. Fortunately mine polished up a treat as did the fork shrouds. For a short time these shrouds were painted black on UK models and some export markets but not on bikes bound for the USA.

HOW TO RESTORE TRIUMPH BONNEVILLE T140

Grab rail/rack

My bike was fitted with the small chrome rack, as standard. It was secured by the top shock absorber bolts, and, at the rear, two 7/16 bolts through the mudguard. It was in good condition, and only required a clean before refitting, which was fortunate, as replacements are far from cheap. When refitting, I made sure that the wiring protection plate was refitted on the underside of the mudguard. It was held in place by the same nuts and bolts that mounted the rack.

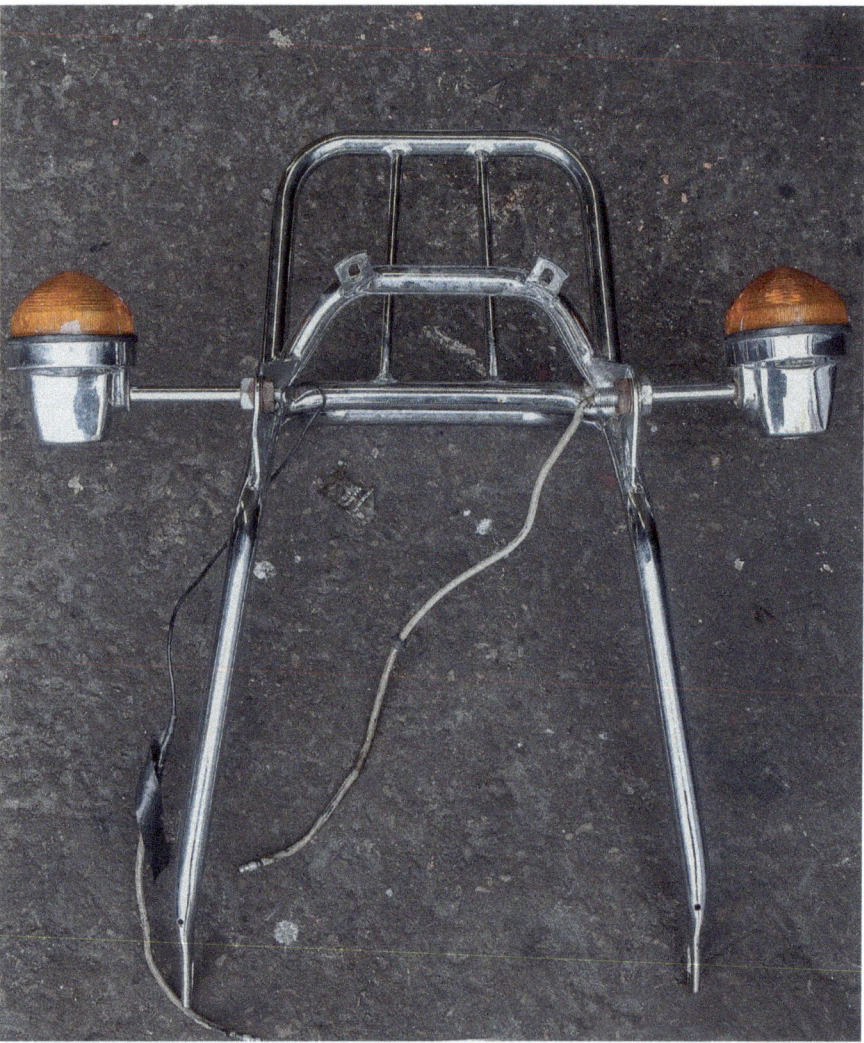

9.69 I was fortunate that virtually all the chrome work on the bike was in good condition, although even bits such as this grab rail can be bought as a reproduction part. Replacing all the shiny bits would have quickly become expensive.

9.70 The cable clips on the bars were tarnished, and the plating was thin, so wouldn't respond to cleaning as well as I had hoped.

9.71 Stainless replacements were sourced. Slightly less shiny than chrome, they were cheaper and would need less looking after in the long run.

GENERAL PLATING

It was extremely likely that smaller fixtures and fittings, including nuts and bolts, would not have survived well; even on my bike, which had not been exposed to European winters, there were quite a few rusty fittings. Having them re-plated was not that expensive, but take everything that needs doing in one go, as lots of platers charge by weight, usually with a minimum charge, so multiple trips would cost more money. Keep a list of exactly what you hand over, as it is reportedly not uncommon for small bits to go missing. Have a chat about the finish you will receive, as Triumph bits have a not unattractive, slightly dull appearance, and some zinc finishes can be a lot brighter. If there are only a few bits that need doing, it might be cheaper to just buy new.

If you like doing everything yourself, then there is also the option of a DIY plating kit. Results can be very good, but are entirely dependent on the time taken to prepare the item, and the level of fine current control, whilst the process took place. Basic kits are cheap; decent ones, with most of the required equipment, less so. Like so many options when restoring, it depends on how many bikes you might end up doing, in order to get the most from your financial outlay.

Stainless is an option, and the list of replacement fittings for the Triumph is huge, but it was a bit too bright for my taste. It is also not

132

TRIM AND BRIGHTWORK

immune to dissimilar metal corrosion, so if inserted into alloy, a barrier of grease should still be employed, to be safe.

9.72 There were very few really rusty fittings on the bike, although this one had suffered badly. Lighter corrosion could be removed and the fitting re-plated.

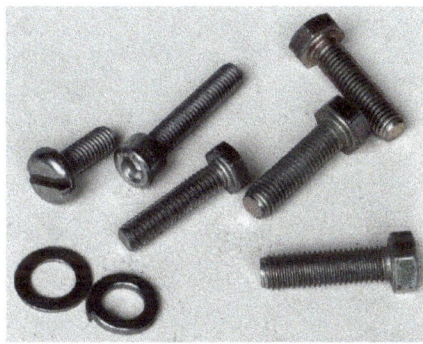

9.73 If a few bits only had needed doing, it would almost certainly have been cheaper to replace with new. Electro-plated items, such as the bolts on the right, above, can be sourced from engineering supply shops at very low prices. Stainless is a popular option, but more expensive. The original dull zinc finish applied at the factory has a quality feel about it, though, so I cleaned and retained as many original fittings as possible.

9.74 Corroded brackets are best dealt with by immersion in one of the rust killers on sale, or you could make up pretty much the same stuff yourself, using a bag of citric acid from a grocery store.

POLISHING

Several alloy bits on the Triumph have a polished finish, and I used a kit of mops which fitted on to a bench grinder, as shown in the Fork chapter. Imperfections, such as scratches, had to be removed using wet-and-dry. I stuck to 1500s and lots of elbow grease for the parts that needed the most work – coarser paper would do the job much quicker, but alloy is soft and it would be easy to do more damage, unless care was taken. This

9.75 Polishing kits are remarkably cheap: this one was £30 at an autojumble, and even came with a spindle to mount it on a bench grinder.

left the surface smooth, but dull and grey, but a quick whizz on the buffing wheel saw a shine return pretty quickly. Once the alloy had been polished, there were two options to keep it looking that way: lacquering, which can be hit and miss, with subsequent flaking being a common problem; or simply using a good polish regularly. I chose the latter. If lacquering, then one designed for high temperature work may adhere better than a straightforward top coat.

9.76 Light marking like this was polished out. Heavier marks needed to rubbed down, first.

9.77 The alloy casting for the rear light was a work of art, far superior to the pressed steel used by many competitors. It was buffed, and responded well.

HOW TO RESTORE TRIUMPH BONNEVILLE T140

RUBBER PARTS

One thing that has been pretty much common to all the restorations that I have done, is the quality of some reproduction rubber parts. Their moulding can often be very accurate, but their longevity is regularly no match for the originals, so don't be in a hurry to dump parts, unless they really do need replacing. Unfortunately, another thing common to all my restorations was that virtually all the rubber bits needed replacing, so you cannot win. However, the footrest rubbers on my bike, for example, only had minor cracking in the depths of their moulded grooves, so they were retained. The centre stand and gear change rubbers had to go, as they were either rotted or missing, and, as expected, their replacements were a bit thinner and softer than the originals; as were the fork gaiters I had fitted. The replacement kickstart rubber was the exception, as it was so sturdy that it was a major undertaking to get it back onto the lever.

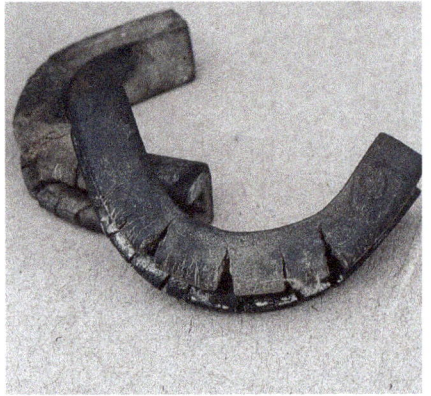

9.78 Old rubber parts were pretty much fossilised on this bike, the downside of the good weather which had preserved the chrome so well. At least rubber bits are cheap.

9.79 The original footrest rubbers had light cracking which looks worse in the picture than in daylight. Stick or twist? I stuck in this case. These items featured in the brochure for my model, referred to as 'contemporary styled footrest rubbers.' I think Triumph was clutching at straws a little by 1979.

9.80 The kickstart rubber was really tight, so I greased the shaft well with rubber grease. Not normally a good idea, but I knew this one wasn't going to slide off again, as it was unbelievably tight.

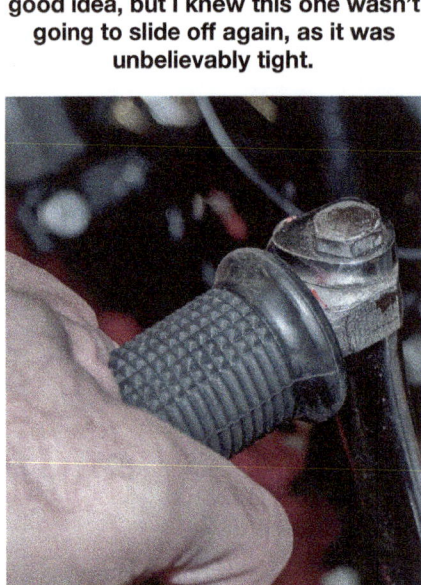

9.81 Even after warming the rubber and greasing the shaft, it was a major struggle to get it in place.

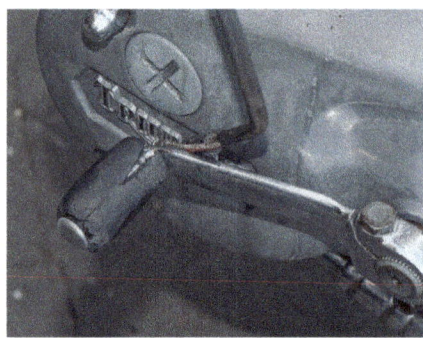

9.82 The gearchange rubber had rotted and, looking at it, it seemed to have been the fully enclosed type.

9.83 The replacement ordered by part number came as the open-ended, sleeved type. I am not sure which would have been fitted originally, but I prefer the look of the rounded enclosed one more, so this might get changed again later.

9.84 The old twist grip rubbers were sliced with a Stanley knife and peeled off.

9.85 There were a lot of glue remnants on the left-hand bar, and also the throttle tube. These were removed with panel wipe.

TRIM AND BRIGHTWORK

9.86 New grips were sourced, and slid on relatively easily, once they had been warmed with a hot air gun.

9.87 The battery was held in place by a rubber strap with a buckle. Neither were on my bike when it was purchased, so new ones were bought. It is a simple but effective method, sadly let down in this case by the replacement rubber strap being too wide to fit through the frame or the slots in the buckle. It was trimmed down with scissors.

BADGES

US tank badges like mine were held by two 2BA screws; the UK version by four, 8-32 UNC. Reproductions of all types are available, and at reasonable prices, which is fortunate as the old ones will probably be suffering from loss of enamel, or pitting. The fit of some of the cheaper copies may leave a little to be desired.

9.88 New badges can be inexpensive. These fitted pretty well, a little out at the front but nothing glaring. Money well spent, as they brightened up the tank hugely.

9.89 It was suggested that the badges were prone to vibrating off, so a small amount of locking compound was added to the screws before fitment.

9.90 The side panel badges were cracked, a common issue …

9.91 … and, having spent time and money painting everything, I couldn't let it down with the old ones, so on with the new.

9.92 The old badges had been held on with a motley collection of screws and bolts. These were the original-type fittings that used spire clips on the rear.

REAR LIGHT/REFLECTORS/LENSES

The rear light unit was cast in alloy, so cleaning and buffing revived that without issue. For 1977, that alloy was painted black. There were two orange reflectors fitted to my bike, which were simply secured inside the neck of the frame with $^{11}/_{32}$ nuts. UK bikes should have them mounted on small steel brackets, which moved them forward slightly. The rear light lens was another item where the decision to replace or retain reared its head, as it had a couple of hairline cracks, which were only visible on very close inspection. I decided to keep things original, although I have a feeling that it should have been stamped 'Lucas England,' so maybe I was only preserving a replacement part anyway.

9.93 Export models had side reflectors, which can suffer from cracking of the plastic lenses. Mine were okay, which was lucky, as replacements are not as cheap as I had expected.

9.94 The rear lens had a couple of very fine hairline cracks. I chose to keep it, as they were really only visible on very close inspection.

Chapter 10
Electrics

Electrics are often the least pleasurable part of a rebuild for many enthusiasts, and I was no different, so my heart sank a little, when I found my project had no fewer than 17 new fuses stashed away under the seat and in the accessory tool bag. Even the most pessimistic of British bike owners would not venture forth with that amount of back up, unless there was a serious problem; it was just a matter of finding it.

From model year 1979 and engine number HA11001 like mine, the electrical system was negative earth. Bikes built before that were positive earth.

10.2 For any in-depth testing, like that set out in the manual, a multimeter was going to be needed. They can be bought very cheaply, but real accuracy costs money.

LOOM

New looms are available for most models of T140 at reasonable prices, and would be the best solution for seriously degraded and /or bodged wiring. Be careful when ordering, though, as this was one of the few

10.3 Several connectors on my old loom were stuck in their sleeves, and the wiring was brittle and snapped during the removal process.

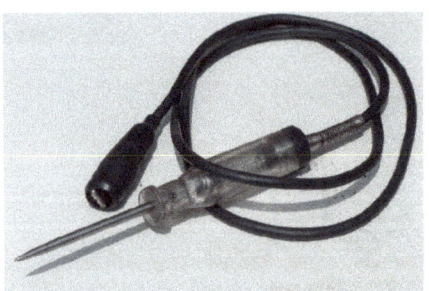

10.1 A simple test lamp would be enough to check that power was reaching components.

10.4 There were also several non-standard repairs and additions. This joint had at least been soldered ...

ELECTRICS

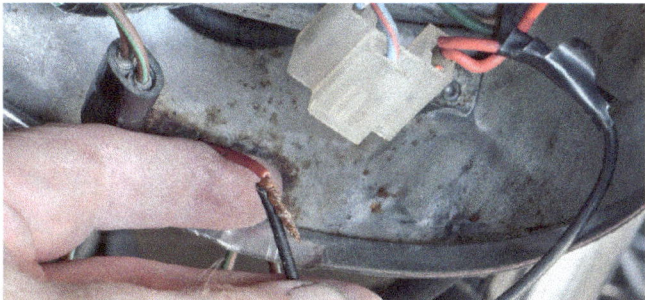

10.5 ... unlike some of the others, which were simply twisted together without any attempt at insulation.

10.6 The wiring in the headlamp was tucked neatly out of the way. I opened it out a bit, and took photographs for later.

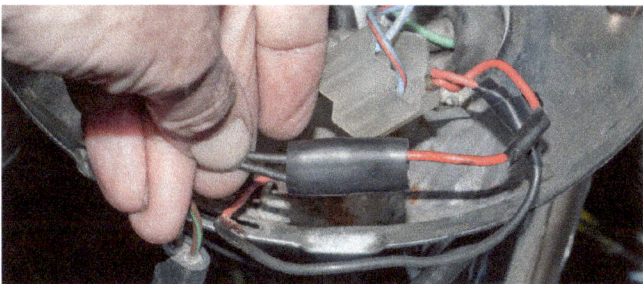

10.7 Some of the wiring to the headlamp was of non-standard and conflicting colours, with more twisted wire visible under my fingers.

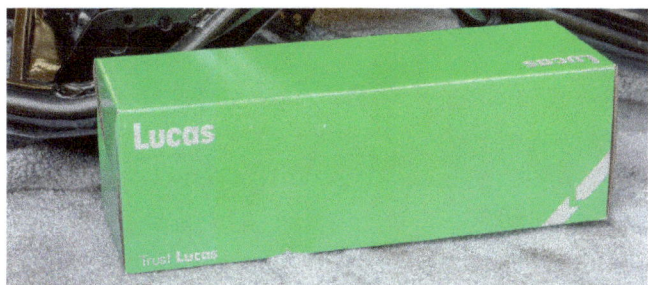

10.8 None of these discoveries was at all unusual on an old bike, but it was definitely time for a new loom. I ordered a Lucas one.

10.9 Out of the box, it appeared to be good quality, but braided, unlike the original. All those wires looked a bit daunting when seen rolled up together. The description on the box, 'T140, single zener, 1979 to 1980,' matched my bike perfectly.

10.10 It came with a really good full colour diagram, so I felt confident of getting it all back together correctly.

10.11 The old loom was held to the frame by these metal straps, which were carefully removed, and cleaned for reuse later.

HOW TO RESTORE TRIUMPH BONNEVILLE T140

10.12 The ignition switch unit had obviously been replaced at some time, and sported non-standard connectors.

10.13 The loom passed through the frame, isolated by these rubber rings. Although badly degraded, they were rock hard and a real pain to shift. The loom was pulled out towards the front, but the wiring was so rigid it, too, was a struggle to release. With hindsight, I should have just cut it rather than mess about.

10.14 The new loom was threaded through the repainted frame in reverse direction to removal. Being flexible, it was easier, but this first section at the frame neck was still slow going as the metal breather pipe lived up there.

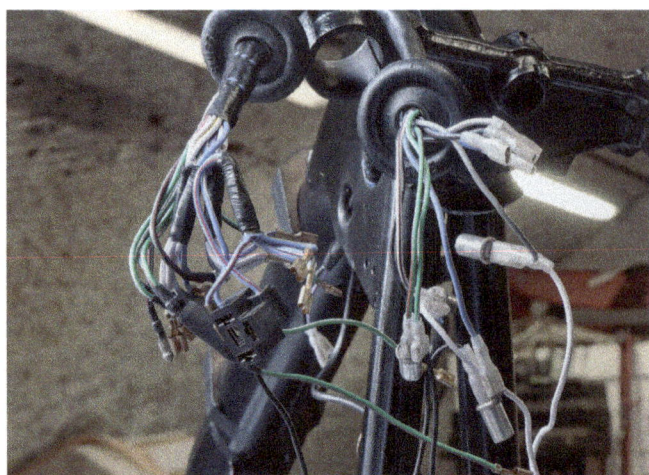

10.15 The loom was loosely laid out at the front, waiting for the headlight and other components to be bolted on later.

10.16 The wiring into the instrument panel was slipped through.

10.17 There were two sections splitting off to feed each pod: lots of fiddly threading, but that was the loom all roughly in place.

138

ELECTRICS

10.18 Having attached it all, I had a quick run round the connectors in case anything was misrouted. I quickly discovered that the connector for the Lucas Rita was missing, and it became obvious that it was the wrong loom despite the description on the leaflet, the box and the seller's website. It all had to come back out, and another loom was sourced. Anyone want a T140 loom, cheap?

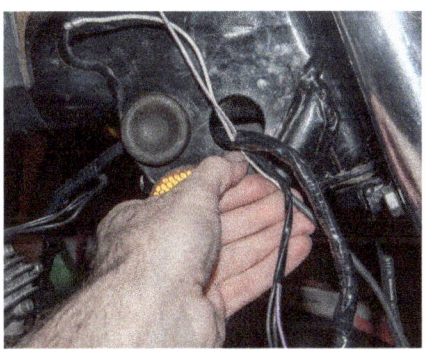

10.19 The second loom was made in the UK (not Lucas, as I could not find a Lucas one that matched the bike; the only other Lucas option was for the later triple zener bikes, which may have worked, but I did not want to chance it). This new one was also good quality, and plastic-sleeved like the original, so it was easier to fit, as it slid past frame obstructions better.

10.20 I also bought a couple of spare rubber grommets, for the wiring from the handlebars into the headlamp shell.

10.21 The old instrument tell-tale lights were separate holders with capped bulbs.

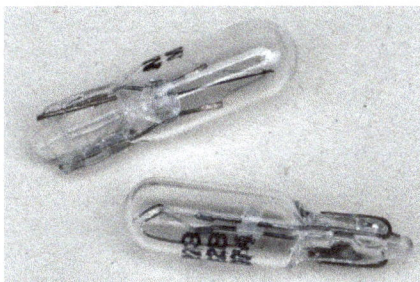

10.22 The new loom came with moulded holders permanently attached, fitted with cap-less bulbs. More modern, and easier to source, but more work, if a holder becomes faulty.

areas where the parts numbers were not infallible, especially for bikes built between 1978 and 1980, as I found out.

The main issue with mine was sticking bullets inside the sleeve connectors – a common problem, making them difficult to separate, which resulted in several breakages. There were also a couple of alterations, where crimped connectors had been used, which did not have a firm grip on the wire. Getting the loom off meant releasing the ignition switch, which should have had spade connectors, but it had been replaced with one which used screw terminals. The rubber top breather pipe should be removed to allow the loom through, and access was made easier when the side reflectors (US) and the horn were removed, as well. The rubber grommets which protected the wiring were a degraded gooey mess. With the loom removed, the wires themselves were revealed to have become hardened and brittle – not unexpected, considering the bike was 37 years old.

WIRING REPAIRS

Even with a new loom, there would still be some sections in the sub-assemblies which do not come with it, that might require some remedial work. Appropriate bullets and connectors are freely available, so the only question was whether to crimp or solder? The latter was without doubt the stronger, but would be more susceptible to vibration, which could be a problem with a Triumph twin. It was also a harder skill to acquire, and an incorrectly soldered joint could be a real problem. For me, though, soldering was the chosen option for the remaining wiring repairs, if crimping, I would have opted for a quality Lucas plier to do the job. Colour-coded wire can be bought easily online, but your local auto electrician should not be dismissed, as it may sell in shorter lengths, and mine was very helpful with advice, to boot. Modern thin-wall wire was smaller in cross-section than the original stuff, but that made it easier to work with and thread through holes. Standard size was always an option for originality.

10.23 Bullets and sleeves can be bought singly, but most suppliers offer multi-packs, which represented better value.

10.24 The bullets were available in either crimp or solder form – or like these: suitable for either method.

HOW TO RESTORE TRIUMPH BONNEVILLE T140

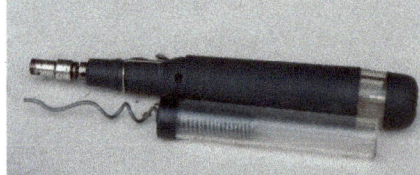

10.25 I chose soldering, as it made a more secure joint, in my opinion. A small gas iron was used. It can be refilled from a cigarette lighter can.

10.26 If crimping had been the choice, then a set of pliers would have been needed. The top two look identical, but the middle one was a cheap copy of the top, and simply didn't work. The bottom one was equally cheap – although basic, at least it was up to the job.

10.27 I had trouble with some of the new connectors, as the inner metal sleeves ranged from pretty tight to ridiculously loose. Irritating, to say the least, and, once again, sourced from an auction site.

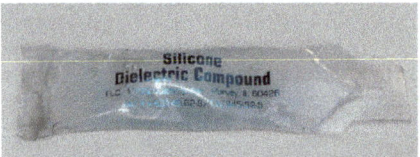

10.28 All the bullets were lubricated with dielectric grease, to discourage the dry corrosion that had plagued the old loom.

10.29 For other wiring repairs, some heat shrink sleeving was bought. A quick waft over with the gas torch or a hot air gun, and it sealed the wiring snugly inside.

ALTERNATOR AND VOLTAGE CONTROL

The Workshop Manual had a comprehensive section outlining the necessary checks for the charging system. Replacement rotors and stators were readily available, and their removal has been covered in the Engine chapter. If more electrical power had been needed, then there were several Lucas bolt-on options. A simple swap from a single- to a three-phase alternator (RM24) would produce more output at lower engine revs. A high output version of this unit is also available, but it would have to be matched to the correct regulator/rectifier, assuming that multiple zeners, like those Triumph fitted, would be too complicated and expensive. The only other possible problem lay in the wiring itself, as I was warned that the original size, connecting the battery to the zener and ignition switch, was marginal when new, so speak to your chosen supplier for its recommendations on the gauge required to carry the proposed amperage.

Zener diode

Much maligned, the zener diode should be perfectly adequate in dealing with voltage control, but suffered if it was not attached firmly to clean metal to provide a good earth. Once again, the post-1980 bikes were beefed up, with three zeners mounted on a common board. A single, new, standard replacement is similar in price to a modern solid state combined regulator/rectifier unit so an upgrade made a lot of sense.

Rectifier

Mounted on the rear mudguard, these are available as a direct replacement part if originality is paramount, but, given the advantages of junking the zener, it made sense to jettison this at the same time.

I went for the combined regulator/rectifier unit mentioned above. There are various makes to choose from, which should all be matched to alternator output.

10.30 The old zener was found bolted to the frame. I suspect this might have been the source of the fuse blowing mentioned earlier, as they are prone to overheating, which then allows excess current to pass, unless they're attached to clean metal and in good air flow – neither of which applied here. For that reason, the factory fitted them to the alloy air box (which had been junked, on my bike).

10.31 If the old rectifier, which is mounted on the rear mudguard, is going to be reused, then a clear photo of the wiring positions should be taken, or the details written down.

ELECTRICS

10.32 The mounting nut was rusty, so I was cautious when removing the rectifier, as the plates can spin, wrecking it internally. Once off, there are various tests that could be done to check it, but I decided it wasn't going back on. If it had been, then the Workshop Manual has all the information.

10.33 The replacement combined rectification and regulation in one fell swoop. It also cost slightly less than a new zener.

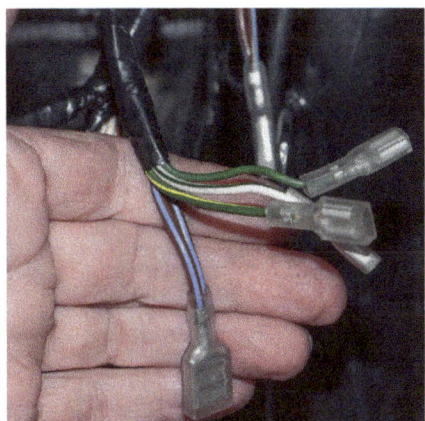

10.34 The new loom obviously came with the correctly coloured wires for a zener installation, but I was slightly nervous about how it would join to the new box when it arrived.

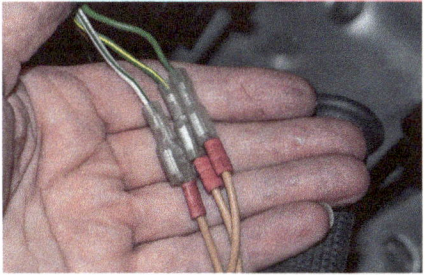

10.35 Fortunately, as a three-phase alternator, the wiring was not critical. Single-phase versions would require more care, so I would have followed the instructions supplied.

10.36 I was undecided where to mount it. Being solid state, it would not get too hot. However, it was finned, so it must emit some heat. I considered fixing it under the seat, but there was minimal air flow there, so I decided to utilise the original hole in the mudguard, and secured it with a bolt and nyloc nut. The contact area should be sufficient to conduct heat away, plus it would get some of the breeze when the bike was moving.

IGNITION SYSTEMS
Electronic ignition
An electronic ignition upgrade for bikes fitted with points seemed like a no-brainer, or at least that was the traditional message. The improvements claimed were: better starting, improved fuel economy, a lack of tedious maintenance, and improved reliability. To some extent, all of that was true, which is why Triumph itself adopted it from 1979. The main negative, though, was that, once it failed, and sadly they do, there was nothing that could be done by the roadside, except to wait for assistance. The balance of risk versus reward would be down to the individual owner and their proposed usage for the bike. For what it's worth, if mine hadn't had Lucas Rita fitted from new, I think I would have gone for an aftermarket electronic upgrade, of which there are several.

10.37 The Rita unit lived inside one of the side panel metal inners. The wiring simply unplugged. The box should always be earthed.

10.38 The units are not bike specific apparently, and I was told that they can be reconditioned successfully. If faulty, though, I think I would have probably gone for a modern full-replacement system.

10.39 The main issue with mine was the wiring at the reluctor, which had pulled from the connector. Both sides of this joint were replaced, but the right-hand one attached to the plate itself, so had to have the old crimped section carefully sliced off, to preserve as much wire as possible; there was not much left to play with.

HOW TO RESTORE TRIUMPH BONNEVILLE T140

Points and condenser

Fitted until 1979 (ish), these were conventional, and their replacement and adjustment is pretty well covered in the Workshop Manual. The condensers were remotely mounted on the coil plate. Several owners mentioned that they had experienced problems with cheap replacement points, so buy from an established specialist, who should be aware of any current problems.

10.42 The points were conventional. Most owners would automatically replace them on a rebuild.

10.40 If you lifted the points cover on an up-to-1979 bike, which included some early E models, this would be what you would find: a twin point setup.

10.43 With the points out of the way, the two lower plates were revealed.

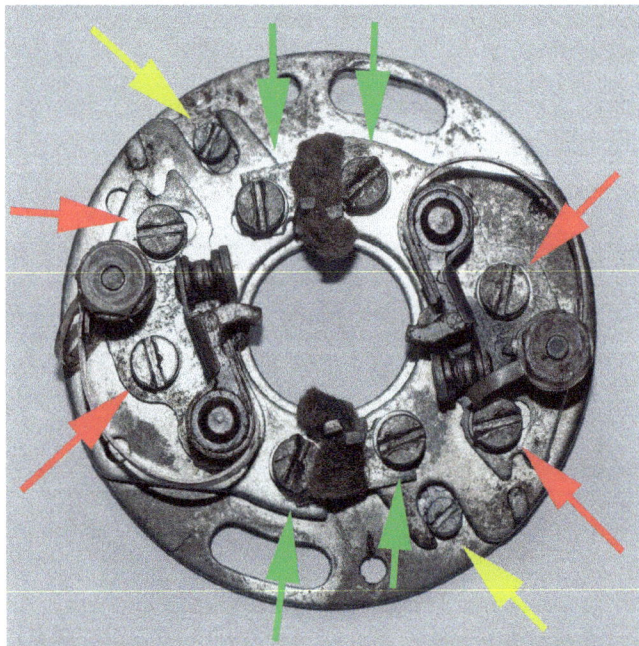

10.41 The manual details it all quite well, but the included line drawing wasn't great. The red arrows show the point locating screws, the green the brush holder and lower plate securing screws, and the yellow the eccentric adjuster screws. Some plates had the eccentrics more centrally-mounted.

10.44 The points cam can wear. Once again, the camera made the marks look worse than they were in reality – this one was good for a few more miles.

ELECTRICS

10.45 On the underside of the cam, the advance springs can be found. These weaken with age, and can cause erratic running when they go soft, so replacement would be a good idea. I was warned that some versions were too strong, and so a mix of one old and one new was required, although I am uncertain whether I would have followed that advice.

10.46 The advance weights should be fine, with lubrication of the pivots being the only thing they would probably need.

Plugs and leads

Pretty much automatic replacement for both of these. Leads could have been checked, as resistance values were included in the manual, but they deteriorate with age as well as use, and replacements won't break the bank.

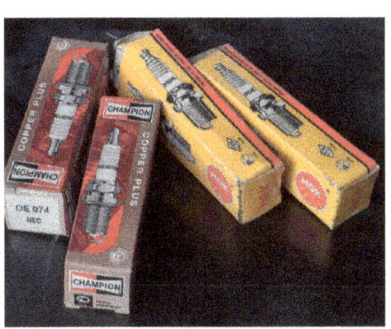

10.47 I have always used NGK plugs in my bikes, but many owners rely on Champion, so I gave them a go. The models with electronic ignition run different plugs from the points bikes.

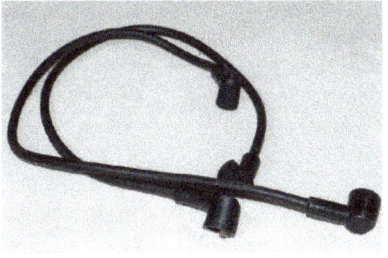

10.48 If you know that the HT leads are old (as were mine), just replace them, as they can be responsible for irritating misfires and hesitancy.

Coils

The coils sat in a plate behind the battery box, isolated from vibration by rubber rings. If you have a points-equipped bike, then they are 12-volt units, Lucas 17M12, wired to fire independently; if electronic, they are a pair of 6v Lucas 17M6, wired in series to fire simultaneously with a dead spark into the exhaust cycle on one cylinder.

Mine initially appeared fine, until I started to clean them, when one crusty patch was found to be covering several deep scrapes and a hole. How they got there, was a mystery, as they were nowhere near any part of the frame, when in their sockets. The damaged one was replaced.

10.49 The coils were held in a separate plate next to the battery box. They were isolated in rubber rings, which were reluctant to let them go after all these years.

10.50 I started to clean them up, and was happy with the finish that was beginning to appear.

10.51 Unfortunately, the second one, (why is it always the second, never the first, one you work on?) had damage, and under one of the scabs lurked a hole.

10.52 A new replacement was purchased, after the other had been checked for resistance and was found to be reusable.

HOW TO RESTORE TRIUMPH BONNEVILLE T140

BATTERY

The battery fitted was a 12N9-4B-1. A lower 7-amp version is often offered by the aftermarket suppliers as a 'suitable' alternative, but Triumph issued a service bulletin warning, which stated that that size would prove inadequate. If you have electronic ignition, either factory-fitted or aftermarket, then it would rely on a well-charged battery to function properly, as I would find out later.

The 1980-onwards models had a new bigger battery, from a Japanese manufacturer, a Yuasa YB14L, to handle the arrival of the electric start. You cannot retro fit this to earlier models, without changing the seat to get the necessary clearance.

This bike arrived without a battery fitted, no doubt removed for transport from the USA. Replacements came in the usual wide price range, but I was a big fan of gel units, having been very impressed by their ability to hold charge over extended periods. They are also good when it came to vibration resistance, which was no bad thing for an old Triumph. My resolve, though, was severely tested, when my local car spares shop offered me a standard one at a very competitive price. All types should be secured by a strap and buckle, both of which were missing, but are cheap to source.

LIGHTS AND INDICATORS

The headlight units pre-1979 model year had the warning lights and the light switch mounted in the shell; later bikes saw the warning lights migrate to a separate panel between the clocks. Lucas replacements are relatively inexpensive; there are also some cheaper aftermarket versions out there, but the quality of a couple I saw was dire to say the least. The headlight on my bike was an aftermarket Japanese Stanley, but fitted with a halogen bulb, a common upgrade. The reflector plays a major part in beam strength and direction, so there would be little point having just a halogen bulb on its own in the old reflector, as a non-matched pairing would do little but add a drain to your electrical system. The conversion had also left my bike without a sidelight, which I found dangling inside the shell. Fortunately, there are lots of options in 7in light units, as it's a common size. I ditched the Stanley, and used a car unit with sidelight, the latter fitted with a high-output led bulb to use as a low-draw, daytime running light.

The indicator units were secured with $^{11}/_{16}$ nuts and, when taking them off, I realised that one of mine was not original. The copy was pretty good shape-wise, but did not carry the Lucas branding on the lens, or the rear of the body. Not something that worried me, but if you are going for a show finish, that would be picked up on by the more knowledgeable, I am sure.

10.53 I had intended going with a gel battery, but this turned up at an unbeatable price.

10.55 The headlamp light unit and rim were secured by a screw.

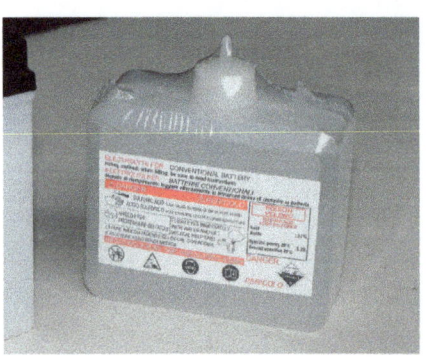

10.54 Packed dry, for transport and longer shelf life, the acid came separately, in the box.

10.56 It should have been Lucas, but had been upgraded to a Japanese halogen unit.

ELECTRICS

10.57 Light output would have been much improved by the upgrade. I was careful not to touch the bulb glass, as it can cause localised overheating and bulb failure.

10.62 Usually indicators fail, as the metal bulb housing and their central electrical contact corrode. These were fine, although the lenses suffered from the usual mashed-up screw heads that secured them.

10.58 If the lamp unit had needed replacing, it was held in place by sprung clips around the edge.

10.60 The indicators simply bolted on, although the securing nuts were really rusty compared to other fittings on the bike.

10.63 The flasher unit was held in a coiled metal holder to provide vibration resistance. It worked; if it hadn't, a new one would have been the only option.

HORN

The horn was simply bolted to a plate under the neck of the frame. It was secured by a pair of $7/16$ nuts and bolts, with rubber washers to give a degree of isolation from vibration. Mine was a non-standard chrome replacement. If original, there would have been a square-ended adjuster screw at the back. If the

10.59 The tail light was conventional, and the contacts were corrosion free. Replacement units would be no problem to source.

10.61 Replica units look authentic, but don't come with any markings. Probably of no importance for many people, myself included, but for a factory-fresh restoration, they would have to go.

10.64 My non-genuine horn was quite attractive in chrome. It should be black, but I wasn't really fussed – it worked, so it stayed.

HOW TO RESTORE TRIUMPH BONNEVILLE T140

unit had been inoperative, turning it 90 degrees left or right whilst pushing the button could have brought it back to life, or perhaps a light tap. Should a replacement be required, an original-style one is around five times the price of a chrome aftermarket one, so it depends how much you are wedded to the correct look.

SWITCHES
There were four types of handlebar switch fitted to T140s, the early type, polished alloy with large paddles and push buttons, then the same but fitted with stickers to label the functions, followed by smaller alloy ones for a couple of years from 1975 before the Japanese inspired versions fitted to bikes from late 1978. Reproductions and rebuild kits were around for the early version but just copies for the latter. Ignition switches were fitted to the left-hand headlamp ear initially then moved to between the clocks. Replacements were no problem although cheap copies do not have marked terminals so would be worth avoiding for that alone in my opinion.

10.67 The other was slightly crustier, but still not bad at all.

10.68 Under the plate, the contacts had been well smeared with red grease. The plastic block was removed carefully, as there was a spring-loaded ball mounted in it, just visible on the right-hand side. Once removed, there was access to the metal contacts.

10.69 Pretty much all classic switch gear was of a similar design, either pushbutton with a sprung ball, or sliding, with a metal sleeve passing over contacts. I just cleaned everything, and re greased.

10.70 The lower section, with the broken lever, was replaced with a new old stock part, as I had trouble finding the lever on its own. Although genuine Lucas, it had been cast incorrectly, and one of the holes had to be opened up before the bolt would pass through.

10.71 The earlier switches were alloy-bodied, and suffered from tarnishing and corrosion – like this one.

10.72 Some also came with a red paddle.

10.65 My bike had the later switches, and had bits missing, as well as the paint loss covered earlier in the book.

10.66 When split, the cover plate on one half was in pretty good condition.

ELECTRICS

10.73 They wrapped around the bars with a vertical split, and were secured with four screws.

10.80 The switches on the left bar were improved with this type, before moving to the black-coated versions. This plain alloy switch was matched with the earlier paddle switch on the right-hand bars.

10.74 The wiring was soldered onto the contacts, set in a plastic block held by two screws.

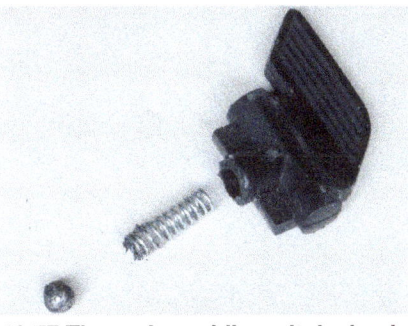

10.77 The main paddle switch simply lifted out of the housing, along with its spring and ball …

10.75 With the plastic block removed, the simplicity of the set up was revealed.

10.78 … as did the two push buttons, each with a pair of springs. All these removable items would come in a Lucas rebuild kit, to rejuvenate the switches, if needed.

ELECTRIC START

From the late Spring of 1980, the Triumph benefitted from the unheard-of modernity of push button starting, thanks to a Lucas TVS starter which was manufactured in India. Unfortunately, it relied on driving the timing gears, which were not up to the job, and the sprag clutch wore out quickly as well. Not many were built, and many were converted back to kickstart only, at the time. The factory quickly brought out a kit of parts to address these issues, and all models from frame number NDA31369 had these improved bits fitted as they rolled off the assembly line. The manual for the appropriate year has a long section on the workings of this rather tardy innovation.

If you own an electric start bike now, and especially one of the few TR65s fitted with them, then their rarity is likely to make them more desirable in future. Finding parts could be tricky, and secondhand bits change hands for very large sums of money, so perhaps it might be financially a good idea to keep using your right leg, and save those pricey and fragile components from any further wear and tear.

10.76 The sprung contacts on the block could then be inspected and cleaned.

10.79 The housing was given a quick scrub on the buffing machine. Heavier corrosion could be rubbed down with wet-and-dry until the surface was smooth, and then buffed, if required. Reassembly was a simple reversal of the strip down.

Chapter 11
Putting it back together

Having worked through the chapters, I now had a collection of restored sub-assemblies, so it was time to put the jigsaw back together. This was the sequence I followed, although a couple of bits have been mentioned previously.

I started with the underside of the frame, and attached the main stand, and put the oil filter in place. In went the swinging arm, the rear shocks, the back mudguard, and the rack. The battery box followed, before moving to the front end, where the yokes went in, but were left unadjusted for the moment. The front forks went on loosely. Then it was time to pause and think about the next steps, or I would end up taking bits off again for access.

The wiring loom was slid through the frame and loosely cable tied in place, and, while I was in the area, the rubber tank mounts were taped in place as well. Next, I fitted the instrument holder; then the top brake hose, which bolted through the yoke; plus the metal pipes and brake light switch block, as both would have been much less accessible once the headlamp went on, which was next on my list. Once in place, I fed the wiring through its various access holes, and squeezed the reluctant new rubber wiring grommets into place.

The rear master cylinder went in next, while there was still plenty of room, as it was tricky to mount, along with its spring and operating shaft.

All the remaining brake parts were hooked up, and the wheels slotted in. It was starting to look good, and it was rolling again, had I needed to move it.

The engine went in, minus the top end, the primary drive, and the timing gear, which made it more

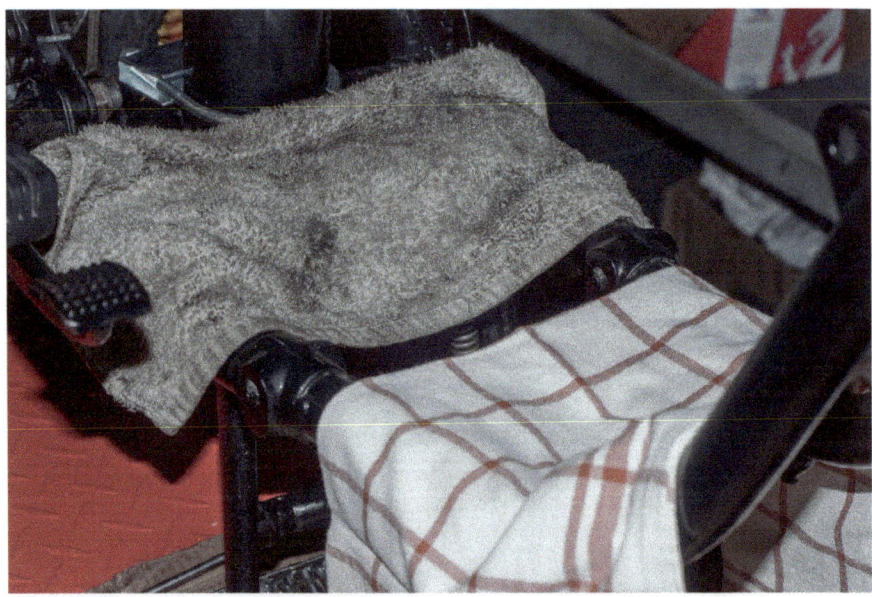

11.1 The frame was protected with towels before I tried to put the engine back, just in case I caught the new paint.

PUTTING IT BACK TOGETHER

11.2 Even in a stripped state, it was a heavy lump to jiggle into place. Lighter than when it came out, though.

11.3 The engine bolts at the front and underneath had spacers, which were of different sizes. The larger one went on the right-hand side.

manageable. The bottom bolt was fitted, first, after a ten minute struggle tapping the spacers into place, and lining them up with a screwdriver, whilst trying to get the bolt through at the same time. The front spacers were also reluctant to seat quickly. The rest of the engine was built up, and the carbs added, making it look very much like a complete bike. I then started at the rear, and worked forwards, connecting up all the loose wiring ends as I went, but leaving everything dangling until I was sure that it was all going to work. The battery was connected up, and all the circuits tested, more of which later.

The front forks were tightened, starting from the bottom – following the instructions in the manual, as there was a set procedure in case of misalignment. The steering head bearings were adjusted, and the pinch bolt tightened.

The frame was filled with oil, and the engine turned over with the plugs out, until I was absolutely certain that it was circulating properly, with regular, strong squirts from the return pipe. The gearbox and primary chaincase were also filled. The brakes were bled; then the seat and tank fitted, and the fuel lines hooked up. So there it sat. The next part should be the best, but was also the most nerve-wracking … would it start?

11.4 All the pipework and wiring was then built back up.

11.5 The carbs were a tight fit over the new rubbers, and needed a bit of wiggling to get them seated properly.

149

HOW TO RESTORE TRIUMPH BONNEVILLE T140

11.6 The rear master cylinder was fiddly and inaccessible, so it went on before any more building-up took place.

11.7 The brake pedal did not want to slide over its square locating shaft without pushing the whole lot back through the frame. The pedal spring was also a pain, but I followed the manual and it eventually went back on.

11.8 The rear brake assembly also proved troublesome, as the mounting plate to the swinging arm could fit in two positions, and it was one area which I had not photographed. The Parts Book was of no help, so it was a matter of fitting the wheel and disc, to determine exactly where it should sit.

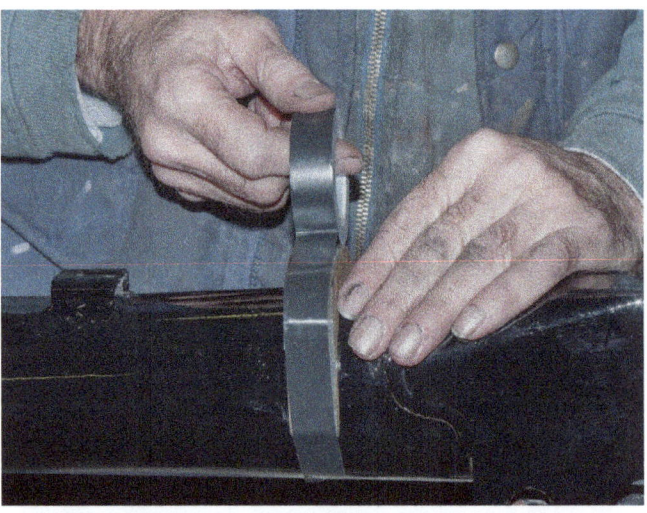

11.9 The wiring was all hooked up, and smaller items, such as the tank rubbers, secured.

11.10 Before starting, the engine was turned over by hand (with the plugs out) until oil could be seen returning to the tank tube in regular squirts.

11.11 It was then checked on the dipstick, and topped up to the maximum level.

PUTTING IT BACK TOGETHER

PROBLEMS

The sequence outlined above made it all seem very straightforward, and the mechanical parts were, really, but it wasn't all that simple: there were quite a few ups-and-downs, before I was in a position to see whether it ran or not.

ELECTRICS

The first issue was with the wiring. For some reason, the wiring diagrams I had did not match my bike. They were close, but unfortunately, that wasn't good enough, so it was back to basics. Colour coding helped, until it came to the unmarked switch, where I had to test the unit for continuity, find the live feed, and work from there. Once it was all hooked up, the neutral light did not work, the indicators did not flash, the rear light didn't work and, by association, neither did the brake lights. There was also no dipped beam. The neutral issue was checked by swapping bulbs, which revealed that one of the brand new ones supplied with the loom was a dud. The indicators were sorted by the addition of an extra earth wire. The tail light was tougher, as both filaments of the bulb were lighting even when one of the two wires were removed. Taking the unit off, once more, revealed damage to the wiring, hidden inside the plastic sleeving, which had allowed cross feeding. Once replaced, there was a tail light – at least, there was after a short, irritating hiatus, as the bulb had blown a filament, unnoticed, during testing. Dipped beam was another faulty bulb, brand new out of the packet, which was pretty unlucky. With everything now functional, I went to test the spark: not a peep, except when the cut-out switch was operated, which gave a fat, but fleeting, burst of blue at the plug. The battery was getting low by now, and I had been struggling for a few hours by this time, so it was left on charge, and I gave up for the day, hoping that the extra boost in the battery might help the ignition.

After an overnight trickle charge, there was enough juice in the battery to fire up the ignition, and I was rewarded by a healthy spark at the plugs. The joy was short-lived, though, as the stop light then became

11.12 The electrics were troublesome, made worse by items such as this top/tail lamp bulb, which had failed during testing, yet still had perfect filaments. Some days in the restoration process can be very trying.

11.13 An intelligent, low-output charger is a necessity with a bike battery. In the past, when I only had a car charger, I used to slave the motorcycle one off a car battery, when charging, to reduce current. One of these smart chargers is much easier and safer.

11.14 To prepare for the start, I set the Rita in the midpoint of its slots, as there were no witness marks visible.

11.15 Once running, I fine tuned using a timing light. My bike had a handy pointer built into the primary case.

intermittent, so the rear brake switch was cleaned and lubricated, and that, thankfully, was that for the electrics.

CARBS

The arrival of a spark meant an attempt to fire up, which it did fairly promptly, but with a lot of coughing and spluttering. The timing was clearly out, by what seemed enough for deep paranoia to set in, and I whipped off the timing cover to make sure that I had indeed managed to line all the cogs up properly. All good there, so another round of blowing out carb orifices, and even trimming the float bowl gasket in case it was fouling the float. Still no improvement, with the blowing back through the carbs making me then suspect a sticking valve, as the new one had been a tight fit in its guide. I let the motor cool, as I sat in my workshop chair contemplating head removal, then busied myself with other small jobs. Much later, I tried again, and this time it fired up a treat, on both cylinders, and it revved out without hesitation. If the valve had been stuck, it had certainly loosened now. A quick check with the timing light, a tweak of the throttle stops, and a round of balancing on the air screws, and it ran really well. No noises, no smoke, just a small amount of tappet tick, as I had set the exhausts two thou looser than spec (I find it helps older engine designs cope better with unleaded fuel – just my preference). So the emotions of the day had swung dramatically from initial despondency to a gentle satisfaction.

CLUTCH

The joy at getting the bike running was tarnished slightly by another issue: the clutch was ridiculously heavy. I thought I had set it all up correctly, so there could only be two possible culprits – either the pressure plate was not lifting evenly, or the clutch springs were the wrong ones. The primary cover came off, once more, and the clutch lever pulled and held back, whilst the engine was turned over. The plate should be parallel to the basket at all times, so it was readjusted by releasing the clutch nuts, as required, a small amount at a time. I still wasn't happy

HOW TO RESTORE TRIUMPH BONNEVILLE T140

11.16 The clutch lift can be gauged by watching this gap, indicated by the screwdriver, which should be even all around the plate. I rotated the engine, which made differences in the gap easier to spot. I adjusted the clutch nuts to get an even lift.

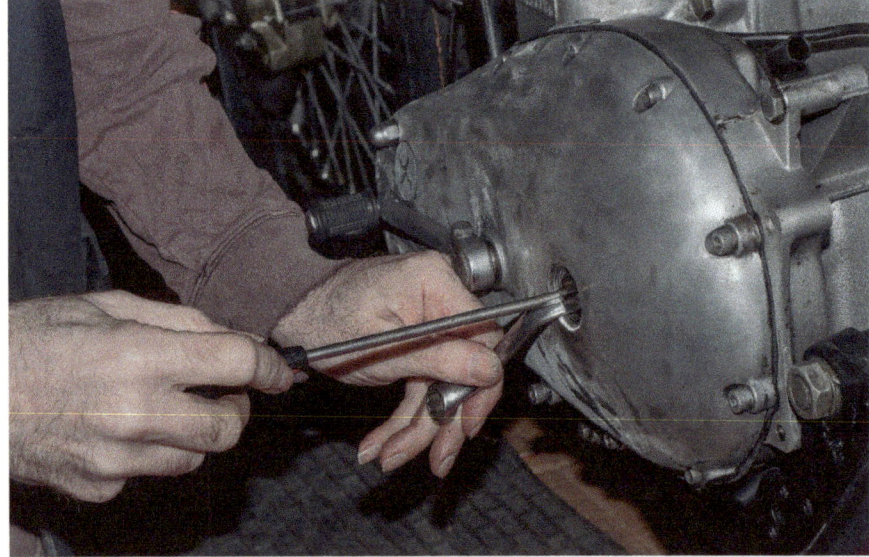

11.17 The clutch pushrod should be adjusted with all tension off the cables. I screwed it in until it just bottomed, then wound out half-a-turn and locked it.

11.18 I used the only motorcycle 20:50 at my local car shop. It was semi-synthetic and would only be in for a few hundred miles. A high zinc formulation would be best for long-term use.

11.19 I have not got round to refitting my oil filter yet, but, when I do, I would go for something like this. No less accessible than the fitting that came on the bike, but upright, so the filter can be filled before fitting. A black one would be hardly visible.

11.20 Gear oil was basic. More modern formulations were suggested to me by other owners, who reported good results.

with the pull weight, so changed the springs as well. I then readjusted everything, staring with the pushrod, with all the cables completely backed off, then took up the slack at the gearbox end adjuster, leaving the outer cable loose by the tiniest amount. This final amount was removed at the bar end, which gave a light pull with a good action, so a success, in the end.

OIL

Not a problem as such, but very much a vexed issue with owners of classic vehicles, and often the source of heated debate. For running in, I used a semi-synthetic, as the full fat version won't let engines bed in very well, plus it would be needlessly expensive, given it was going to be changed very quickly. I used a modern 20:50 blend designed for

PUTTING IT BACK TOGETHER

Harleys, although many swear by classic versions, which have even lower detergents and higher zinc content. These would be a good choice, where a separate oil filter had not been fitted, if you have one on the bike, then detergency would be an advantage. I intended in future to use a medium-priced, high-zinc oil, changing it frequently, to give the best balance between the demands on my pocket and protecting my bike. Other opinions are available. In the gearbox, I used a 80W/90 GL4, non-hypoid: outdated without a doubt, but kind to the bushes within. Many owners strongly recommended the use of Redline Heavyweight Gear Oil, so that may be an option later.

WAS IT WORTH IT?

This bike gave a pretty fair representation of the issues often found with a low mileage repatriation. It had not required any heavy-duty engine work, and the chrome was pretty good, but, despite those major advantages, the final cost was around the price that would have bought me a Bonneville ready-to-go from a classic bike dealer. In the absence of a clear cut financial advantage, solace could be found in the fact that I knew that every part had at the very least been stripped, cleaned and inspected, so there should be no unexpected surprises lurking in the near future. The most

11.22 From whichever angle I looked at it, the Triumph was a handsome-looking bike.

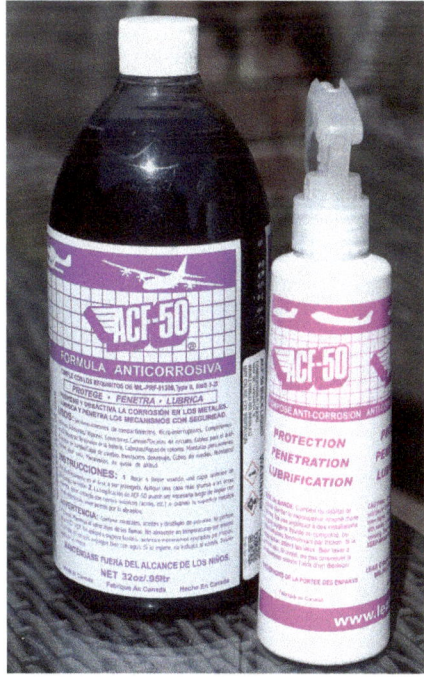

11.21 The bike was thoroughly cleaned and polished, then coated in ACF-50 to provide corrosion protection. Once again, a product recommended to me by many other classic enthusiasts.

153

HOW TO RESTORE TRIUMPH BONNEVILLE T140

11.23 The cases had cleaned well. Vapour blasting would have been easier and quicker, but these were perfectly acceptable ...

11.25 Even the pin-striping had a factory look.

11.24 ... as was the tank. Not up to Meriden standards, but clean and tidy.

11.26 I was a happy man. My first foray into the world of British bikes had been a rewarding and enjoyable experience. Time for one of these, whilst I pondered the potential pleasures of a whole new biking world.

important bonus, though, was the satisfaction that I had done it all myself, and the process had been immensely enjoyable. Before the off, I had wondered if it would be a good introduction to the world of British motorcycles, and I certainly had not been disappointed.

Also from Veloce Publishing –

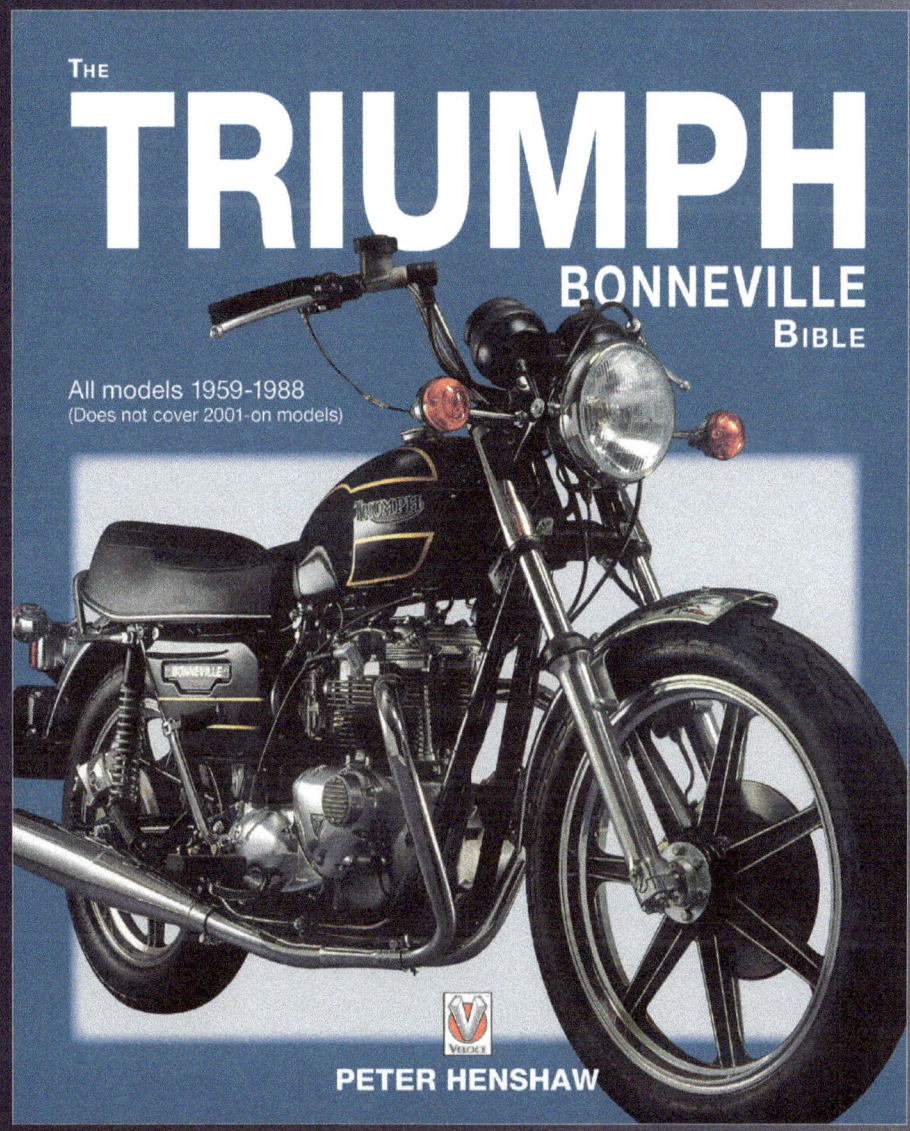

The story of the Triumph Bonneville – its conception, design and production, how it compared to the competition (British and Japanese), and how it was seen at the time. With insights into the company that built it, from the boom times of the 1960s, through struggles in the 70s, and eventual closure in the '80s, plus guidance on buying a Bonneville secondhand, this is the fascinating history of a British icon.

ISBN: 978-1-845843-98-4
Hardback • 25x20.7cm • 160 pages • 262 pictures

For more information and price details, visit our website at www.veloce.co.uk • email: info@veloce.co.uk • Tel: +44(0)1305 260068

Also from Veloce Publishing –

Long time Meriden worker and Triumph restorer, Hughie Hancox, describes everyday life in the Triumph Production Testing team from 1960 to 1962. A story packed with amusing anecdotes, and guidance on fixing problems still found today on the 1960s models. An intimate and entertaining account of Britain's most famous motorcycle factory.

ISBN: 978-1-845844-41-7
Paperback • 25x20.7cm • 160 pages • 183 colour and b&w pictures

For more information and price details, visit our website at www.veloce.co.uk
• email: info@veloce.co.uk • Tel: +44(0)1305 260068

A fascinating record of life in the Triumph motorcycle factory.

ISBN: 978-1-787115-49-1
Paperback • 25x20.7cm • 144 pages • 91 b&w pictures

For more information and price details, visit our website at www.veloce.co.uk
• email: info@veloce.co.uk • Tel: +44(0)1305 260068

Also from Veloce Publishing –

A practical, straightforward guide to buying a secondhand Triumph Bonneville, from the very first 1959 T120 pre-unit 650, to the very last T140 unit 750 machines built by L F Harris. What they're like to live with, spares availability and prices, plus point-by-point guide to buying a Bonnie. One hundred colour photos, useful appendices and expert advice mean this book could save you thousands.

ISBN: 978-1-84584-134-8
Paperback • 19.5x13.9cm • 64 pages • 127 colour pictures

For more information and price details, visit our website at www.veloce.co.uk
• email: info@veloce.co.uk • Tel: +44(0)1305 260068

ISBN: 978-1-845846-09-1
Paperback • 19.5x13.9cm • 64 pages
• 104 colour pictures

ISBN: 978-1-845847-55-5
Paperback • 19.5x13.9cm • 64 pages
• 102 colour pictures

Having these book in your pocket is just like having a real marque expert by your side. Benefit from the author's years of Triumph ownership, learn how to spot a bad bike quickly, and how to assess a promising bike like a professional. Get the right bike at the right price!

For more information and price details, visit our website at www.veloce.co.uk
• email: info@veloce.co.uk • Tel: +44(0)1305 260068

Index

Alternator 68, 69, 92

Badges 135
Barrels 54, 55, 64-66, 94
Base coat 118
Battery 144
Battery box 22
Bearings steering head 23-27
Brake bleeding 50
Brake calipers 43-47
Brake discs 47
Brake hoses 49
Brake master cylinders 38-43
Buffing 121

Carburettor
  Amal Mk1 99-103
  Amal Mk2 103-108
  Bing 109
Clutch 70-72, 92-93
Coil plate 22
Coils 143
Crankshaft 85-87
Crankshaft bearings 85-87
Crankshaft pinion 74, 75
Crankshaft sludge trap 86
Cylinder head 51-53, 58-62, 95

Engine removal 67
Exhaust system 109, 110

Filler 116, 117
Fork gaiters 34
Fork seals 31, 32

Frame 16-19
Frame number 19
Fuel tank 96-98
Fuel tap 97

Gearbox 76-80, 90, 91
Gearbox quadrant 81, 82
Grab rail 132

Handlebars 22, 23
Headlight 144, 145
High gear assembly 89, 90

Ignition, electronic 74
Ignition points 142, 143
Indicators
Inner tubes 112
Instruments 129, 130

Jets 101, 107

Kickstart assembly 80, 81

Lights 144, 145
Lucas Rita 141

Mag wheels 111
Mudguards 131
Multimeter 136

Oil 152
Oil filter 21
Oil pump 75

Paint removal 17, 115
Piston and rings 55, 63
Powder coating 18
Primer 118
Pushrods 53, 95

Regulator 140, 141
Rocker assembly 56, 57
Rubber parts 134, 135

Safety, workshop 10
Seat cover 125-128
Shock absorbers, rear 34-37
Sparkplugs 143
Stands 21, 22
Steering lock 27
Studs, engine 57
Swinging arm 20
Switchgear (painting) 123
Switchgear (repair) 146, 147

Tappet blocks 64-66
Tools, hand 9
Transfers 124, 125
Tyres 112

Valves and guides 58-62

Wheel bearings 113
Wheels 111
Wiring loom 137, 139

Zener diode 140

www.ingramcontent.com/pod-product-compliance
Lightning Source LLC
Chambersburg PA
CBHW040739300426
44111CB00026B/2988